GENE BANKS AND THE WORLD'S FOOD

GENE BANKS AND THE WORLD'S FOOD

DONALD L. PLUCKNETT,

NIGEL J.H. SMITH,

J. T. WILLIAMS,

AND

N. MURTHI ANISHETTY

PRINCETON UNIVERSITY PRESS

PRINCETON, NEW JERSEY

Copyright © 1987 by Princeton University Press

Published by Princeton University Press, 41 William Street,
Princeton, New Jersey 08540
In the United Kingdom: Princeton University Press, Guildford, Surrey

All Rights Reserved

Library of Congress Cataloging in Publication Data will be found
on the last printed page of this book

ISBN 0-691-08438-6

This book has been composed in Linotron Baskerville

Clothbound editions of Princeton University Press books are
printed on acid-free paper, and binding materials
are chosen for strength and durability

Printed in the United States of America by Princeton
University Press, Princeton, New Jersey

To the memory of

NIKOLAI VAVILOV

Plant explorer, geneticist,
and biogeographer

CONTENTS

PREFACE

Gene banks have arrived on the center stage of scientific and public-policy debates. Intense public interest in genetic engineering has raised expectations that biotechnology will unveil dramatic improvements in crop plants and domestic animals. While breakthroughs in genetic engineering are occurring with increasing speed, options for the future are being foreclosed by the erosion of one of the world's most important heritages, the genetic diversity of our crop plants and their wild relatives. On every continent and even on small island nations, crop gene banks, where seeds are kept at reduced temperature and moisture and where other plant materials are kept in test tubes or in field collections, have emerged as linchpins in the global effort to conserve as much of the gene pool of crop plants as possible and to tap this reservoir for the benefit of mankind.

Although there is general agreement, both in the scientific community and among the public in industrial and developing countries, that conservation of plant and animal resources for agriculture and other purposes is warranted, questions have arisen about the best way to preserve genetic diversity. Some argue, for example, that old varieties of crops should continue to be grown even after they have been discarded by farmers. Others insist that for most crops, gene banks are the most practical way to safeguard crop genetic material while related species are conserved in their natural habitats. The scientific community has opted strongly in favor of gene banks allied with associated and complementary methods of conservation.

Differences of opinion also abound about who should own and operate gene banks, with some observers contending that crop germplasm collections are controlled mostly by Western industrial countries and, by inference, for the benefit of multinational corporations. Conservation and use of crop genetic resources have also become enmeshed in a related issue, that of plant-variety rights or patents. Unimproved germplasm of food crops generally moves freely between nations, but some people see patents as a potential threat to farmers.

Gene banks underpin the productivity of much of the farm land in industrial and developing countries, and with the lively discussions taking place in scholarly literature and the popular press, a review of the scientific issues and policy implications of crop germplasm work is timely. This book is targeted for a broad audience of interested citizens, policy makers, researchers in the agricultural and environmental sciences, as well as social scientists concerned with farming and rural development. Our global approach will stress the scientific linkages between developed nations and the Third World and their interdependence in the worldwide effort to boost agricultural production.

The first chapter, "Gene Banks: A Global Resource," provides the rationale for gene banks by outlining the dimensions of the world's growing population and the need to further boost agricultural productivity. The chapter also underscores the importance of gene banks by briefly describing the scope and reasons for the progressive loss of genetic diversity of crops and related species. In this introduction, we briefly describe some of the infamous and lesser-known cases of productivity crashes due to the genetic simplification of farm lands. In Chapter 2, "Seeds in Due Season," we explore the dynamic nature of modern farming by focusing on the rapid turnover or replacement of varieties. We also discuss the breeding strategies employed to make varieties more resilient to environmental challenges. We examine seed-production systems, including hybrid seed, and outline quality control measures, and discuss the actual and potential impact of plant patents.

The history of germplasm preservation and exchange, from botanical gardens to modern cold-storage units, is explored in the third chapter, "Plant Collectors and Gene Banks." The activities of amateur and professional collectors are reviewed and the earlier roles of consuls and colonial services are outlined. The transition to modern gene banks in the twentieth century is also described.

Principles guiding the modern collection and utilization of crop germplasm are covered in the fourth chapter, "Gene Banks." We describe a typical gene bank, including the preparation of material for entry into collections and the methods of storage and evaluation. We discuss some of the scientific issues concerning the most suitable ways to preserve germplasm, including conservation of natural areas for wild relatives of crops, and we analyze problems sometimes encountered at gene banks and give remedial measures.

In Chapter 5, "Biotechnology and Genetic Resources," we outline recent advances in recombinant DNA technologies as they apply to agriculture. Although traditional crop-breeding techniques will continue to be the most important means of incorporating desirable genes into crops for the foreseeable future, some aspects of biotechnology, such as tissue culture, have been employed by crop breeders for decades. We discuss these and other techniques that are being explored by private companies and public institutions and that may soon provide tangible benefits for agriculture. As further progress is made in bio-technology research, the value of gene banks will be enhanced.

The status and locations of gene banks are the focus of Chapter 6, "Genes in the Bank," where we present the latest available information on germplasm holdings by crop. We discuss cereals, root crops, pulses, vegetables, and some industrial crops, and include data on domesticated plants and wild relatives whenever possible, since the latter, besides having other attributes, are often useful in upgrading pest and disease resistance in crops. We note methods of storage, sources of material, and degree of redundancy in collections. Lists of accessions reveal the extent of germplasm collections, and when accessions have been evaluated, can help pinpoint gaps. This chapter shows that Third World countries house a substantial portion of the world's crop germplasm in long-term storage and that gene banks are operated mostly by national programs in both industrial and developing countries, as well as by international agricultural research centers. Private companies concentrate on assembling working collections rather than on operating long-term storage facilities.

The bottom line in germplasm conservation is its effectiveness in serving agriculture now and in the future. Simply storing material, as in a museum, is not enough, and vague promises about potential rewards in the distant future rarely generate much financial support. Policy makers and donor organizations are particularly interested in immediate payoffs in gene banking for farmers and consumers. If gene banks are already serving a useful purpose, current successes will whet the appetite and sustain funding. In Chapter 7, "Gene Bank Dividends," we spotlight specific ways in which scientists draw material from germplasm collections to meet challenges to agricultural productivity, and we give examples of how gene banks have been tapped to upgrade pest and disease resistance and to mold crops to adverse soils and weather.

For the most part, gene banks store samples of traditional varieties and more recent varieties that are no longer in use, but wild species are also represented and play a crucial role in broadening the genetic base of crops. The importance of wild species related to cultivated plants is thus highlighted in Chapter 8, "Wild Species: The Wider Gene Pool." Several of our crops started out as weeds, and some cultivated plants have benefited from spontaneously exchanging genes with wild relatives. The use of wild germplasm in crop breeding is summarized here. The value of wild accessions in gene banks is underscored, as is the importance of conserving natural habitats.

Chapter 9, "A Case Study in Rice Germplasm: IR36," presents a close-up of how plant breeders and other agricultural scientists have collaborated in using rice genetic stocks to produce a successful high-yielding variety, one of a series in the "green revolution." Here we emphasize the dynamic nature of agro-ecosystems and how scientists have responded to the kaleidoscopic array of pests and diseases by effectively using rice germplasm collections to breed new varieties. The IR36 story also vividly depicts how a team approach, uniting the efforts of scientists in various disciplines, produced a viable technology for small-scale farmers. The green revolution in rice began with IR8 in 1966; within a few years, however, this dwarf variety and a few of its successors suffered severe losses in certain areas, particularly in Indonesia, primarily because of damage caused by a rapidly evolving cast of pests and pathogens. With IR36, breeders managed to incorporate resistance to a broad range of diseases and pests that had damaged IR8. The genealogy of IR36 is complex and demonstrates how numerous crop lines, often from several different countries, are involved in the development of modern varieties. A sustained team effort and the availability of valuable genes gathered from several nations and stored in gene banks enabled breeders to launch the most widely planted rice variety in history.

In the final chapter, "Global Imperatives," we summarize how far germplasm conservation has come, and we outline future tasks such as better evaluation of existing accessions, more duplication of collections to prevent irreparable losses, increased training opportunities for germplasm specialists, and further strengthening of national agricultural research programs in the Third World. We discuss the relative merits of strengthening the existing systems of germplasm storage and exchange, or of completely overhauling the current setup. We propose an international fund for genetic resources work, and, finally, we speculate where gene banking is likely to be in the twenty-first century.

March 1986

A NOTE ON TERMINOLOGY

As much as possible, we refrain from using highly technical language in this book, but occasionally we find it necessary to use terms that may be unfamiliar to the reader. In this section we will define certain biological terms for those who are not directly involved in germplasm work.

Gene banks can be either *ex situ*, where seeds or plant parts are preserved outside their area of growth, or *in situ*, where plants, including wild relatives of crops, are maintained in natural preserves. Gene banks contain plants preserved in seed form or in outside plots, known as *field gene banks*. Research is being conducted on storing some plants in tissue-culture form in glass containers (*in vitro*) to save space and reduce costs. And to find a means to store vegetative materials for extended periods, *cryopreservation* experiments are being conducted in which tissue cultures are held at $-196°$ C. Some specialized gene banks also house collections of genetic stocks, such as mutants.

Plants preserved in seed form belong to two classes: those with *orthodox* seeds that can be dried to moisture levels between 4 and 6 percent and then kept at temperatures as low as $-20°$ C, and those with *recalcitrant* seeds that do not survive drying and freezing. Crops that do not produce seed, or those with recalcitrant seeds, are typically stored in field gene banks.

Genes, contained in living organisms, are the information blueprint for all biological life and are responsible for the characteristics of plants, animals, and microbes. A *genotype*, then, is the distinct and unique combination of genes in an organism, and gene banks are currently the only place where plant genotypes are systematically stored. Plant germplasm collections are assemblies of genotypes representing primitive varieties, or *landraces*, of indigenous agriculture, obsolete *cultivars*, or modern varieties that are the product of scientific breeding, and wild relatives, including weedy plants, of crops. We use *variety* and *cultivar* synonymously.

The *gene pool* of a crop is a broad category encompassing the genetic resources of the species, including material that can be crossed with it and that contributes genes. A crop gene pool frequently includes wild relatives. *Gene flow* refers to the exchange of genes between individual plants and between plant populations. *Genetic erosion* is the loss of genes from a gene pool due to the elimination of populations because of such factors as the adoption of modern varieties and land clearing.

A *gene-bank accession* is a sample that has been received for processing and eventual storage and evaluation. It is akin to a library accession that is destined for cataloguing and shelving. To be useful to breeders, accessions need to be screened for their reactions to various *pathogens* (disease-causing agents) and other environmental stresses. Genes from an evaluated accession may make their way into breeding lines that eventually lead to the development of varieties for release to farmers. During the breeding process, it may be necessary to *backcross* the accession with parental lines (usually advanced breeding materials) numerous times in order to retain the desired gene or genes while eliminating unwanted characteristics. Scientists may also resort to *widecrossing*, or breeding crops with other species such as wild relatives, in order to obtain desirable traits. A major goal of plant breeding is *polygenic resistance* to pests and diseases, whereby the variety is protected by several genes. Polygenic resistance usually slows the emergence of insect pests and pathogens that can overcome the crop's defenses. *Monogenic* (single gene) *resistance*, often referred to as vertical resistance, is usually less stable than polygenic resistance.

Accessions at gene banks are usually landraces or traditional varieties selected by farmers. Many gene banks also contain modern varieties no longer in use as well as wild species.

A list of acronyms used throughout the text and the organizations they represent can be found in Appendix 2.

ACKNOWLEDGMENTS

We are grateful to Dana Dalrymple, David Jewell, Quentin Jones, Garrison Wilkes, and an anonymous reviewer for commenting on an earlier version of the manuscript. T. T. Chang and Brent Ingram also made helpful inputs on parts of a preliminary draft. We would also like to thank Judith May of Princeton University Press for her many helpful suggestions for improving the organization and clarity of the manuscript.

Research for this book was partly supported by a Guggenheim Fellowship to Nigel J. H. Smith.

GENE BANKS AND THE WORLD'S FOOD

GENE BANKS: A GLOBAL RESOURCE

The world's human population has reached 4.8 billion and is expected at least to double before it stabilizes. How to feed our growing ranks is a formidable challenge for all mankind. For some countries there is no alternative to increased use of marginal lands for production, and this approach in turn requires crop plants bred for such unfavorable conditions. But if we are to avoid further damage to marginal environments and a downward spiral of ever-declining yields, much of the increased food output will have to come from improved productivity on existing farm lands. Genetic manipulation of plants is one of the most important avenues for improving agricultural productivity (Ayensu, 1978). The only alternative is to open up new farm land in areas far from population centers, and this strategy could destroy or seriously jeopardize natural habitats throughout the world. Our remaining wild areas, storehouses of potentially useful genes for agriculture, medicine, and industry, are increasingly threatened by encroaching settlement.

Boosting and sustaining agricultural productivity, the sane alternative to a further deterioration of remaining wild areas and marginal zones, embraces scientific, social, and political concerns. Improved crop performance relies on sound scientific knowledge in a number of areas, including agronomy, entomology, genetics, plant pathology, and soil sciences. Plant breeding, an outgrowth of genetics, has a central role in the worldwide effort to improve agricultural output, and breeders rely on genetic resources to produce better-adapted and higher-yielding varieties. Maintaining the genetic diversity of crops as well as conserving wild plants and animals has thus become a central principle in strategies for sustainable agricultural development. Scientists, and increasingly the general public, have come to realize the long-term benefits provided by the conservation of biological diversity and the habitats in which it is found.

Since antiquity, farmers have been the custodians of crop genetic resources, but today crop germplasm is increasingly being preserved in gene banks. It is this radical departure from tradition that is partly responsible for some of the controversy over the preservation and use of crop genetic diversity. Gene banks contain germplasm samples within easy reach of plant breeders. Scientists need well-preserved and evaluated materials at hand so that they can confront the many threats to agricultural productivity.

Without the convenience and reliability of gene banks, breeders would continuously have to organize expeditions in search of samples for their breeding programs. Furthermore, gene banks contain traditional varieties that are no longer cultivated as well as populations of wild relatives of crops that might otherwise have become extinct due to agricultural expansion or other forms of development. Wild species preserved in natural habitats must complement *ex situ* gene banks.

The debate and excitement over crop genetic resources have also spilled over into the political arena. Questions have arisen as to who owns genes, and whether it is possible or advisable to patent them. Some individuals argue that collections of the genetic diversity of crops are in the hands of industrial nations and are being exploited mostly by multinational corporations. The industrial countries are perceived by some as the "winners" in marshalling the lion's share of plant genetic resources at the expense of developing nations.

We will document that plant genetic resources are being used by breeders to the benefit of farmers and consumers in virtually every nation. Further, we will demonstrate that the ongoing effort to conserve crop genetic diversity is worldwide, encompassing international, regional, and national institutions, and that public and private organizations tap germplasm collections to produce improved varieties. In addition, we argue that, whenever possible, crop germplasm should be conserved as seeds at low temperatures, or in the case of crops that cannot be stored in this way, in test tubes and in field gene banks. Finally, we stress that the genetic diversity of wild relatives of many crops also needs to be safeguarded in natural reserves.

To foster a better appreciation of the urgency of conserving crop genetic diversity both in gene banks and in wild habitats, we review the origins of crop genetic enrichment, the decline of genetic diversity of farm lands, and the dangers of genetic erosion. Our main purpose in this chapter is to provide the rationale for conserving the genetic diversity of crop plants by exploring cases where agricultural areas have become vulnerable to major downturns in productivity because of a narrow genetic base.

THE ENRICHMENT OF CROP GENETIC RESOURCES

Since the dawn of agriculture at least 10,000 years ago, farmers have been selecting crops suited to a wide range of environments. Starting from a small area, or in some cases several areas, for each crop, plants have been chosen to fit a diverse range of environments and to fulfil dif-

fering tastes, color preferences, aromas, textures, and cooking qualities, among other features. Maize (*Zea mays*),[1] for example, was domesticated in Mexico, and by the time Europeans set foot on the shores of
the New World, the tall graceful cereal rustled in the ephemeral breezes
of high mountain valleys in the Andes, stood tall above competing
weeds in the hot, moist jungles of the Amazon lowlands, and ripened in
irrigated fields claimed from the coastal desert of western South America.

Starting from a finger-sized cob with tiny, popcornlike kernels, farmers have selected races of maize with an extraordinary range of cob
sizes, kernel shapes, and colors. Dried Indian corn, sold in stores in the
United States particularly around Thanksgiving, gives a hint of the genetic diversity of the ancient crop. Some maizes were chosen primarily
to make flour, others to eat after boiling or roasting, still others to prepare beverages. In the Peruvian highlands, for example, a shiny purple
maize is crushed and strained to make a grapejuice-colored and slightly
fermented drink called *chicha de jora*.

As domesticated plants diffused from their origins, men and women
kept a keen eye open for potentially useful traits. When crops were
taken to new islands and continents, their genetic diversity often received a further boost because of new evolutionary opportunities.
When a species is introduced to a novel environment, it often changes
especially quickly when faced with fresh environmental challenges.
New gene combinations are drawn out as crops undergo adaptive radiation and as genotypes respond to different climates and soils and are
attacked by new pests and pathogens (Chang, 1983a). Also, when crops
are taken from their source areas they may encounter different relatives and cross with them. In West Africa, for example, the common
Asian rice (*Oryza sativa*) has hybridized with *O. glaberrima*, an indigenous
cultivated rice, thereby enriching the gene pool of rice in the region
(Ng, 1979; Ng et al., 1983). Sometimes crops develop secondary centers
of variation with greater genetic richness than their source areas (Harlan, 1972).

Farmers have taken advantage of this accelerated process of change
to select traditional varieties or landraces for new micro-environments
(Figures 1.1, 1.2). Under domestication, plants underwent rapid and
radical changes because people exerted strong selection pressure on a
rich pool of genetic variation (Vavilov, 1949). Repeated planting and
harvesting cycles resulted in a buildup of mutations and selection of desirable traits (Harlan, 1965, 1975a; Chang, 1976a). In primitive agri-

[1] The first time a species is mentioned, the scientific name is given. Only the common name
is used thereafter, except in the tables and illustrations.

5

1.1. Two rice (*Oryza sativa*) varieties grown in swamps in the vicinity of Banjar-masin, South Kalimantan (Borneo), Indonesia, February 1985. The low-lying area of South Kalimantan contains thousands of rice varieties developed for different tastes and adapted to numerous micro-environments.

culture, farmers often grow mixtures of genotypes, thereby providing opportunities for further crossing. In cooler, highland areas of Latin America, for example, farmers often plant a multicolored mixture of bean (*Phaseolus* spp.) varieties as a hedge against inclement weather. The different beans have uneven germination and some plants will survive if the early summer rains fail or are late (Clawson, 1985). By the time of Christ, most of our crops had been domesticated and dispersed far from their ancestral homes (C. O. Sauer, 1969). As crops were picked up by different cultures and encountered new growing conditions, they diversified.

The diversification of crops continued during colonial periods, from the spread of Roman civilization until the British, Dutch, Portuguese, Spanish, and others settled in the tropics and subtropics. Expanding cultures exchanged plants between continents and archipelagos. By 1505, for example, Portuguese explorers had taken sweet potato (*Ipomoea batatas*) from Brazil to Goa in India; from there other traders carried the vining root crop to Indonesia and Polynesia (Baker, 1970a:52). By the mid-sixteenth century, sweet potato was a popular garden crop in Spain and Portugal (McAlister, 1984:469).

Columbus brought seeds with him on his first voyage to the Americas, but they apparently perished. On the second voyage in 1492, however, he brought seeds of wheat, chickpea (*Cicer arietinum*), melons, onions, radish (*Raphanus sativus*), salad greens, grape vines, sugar cane (*Saccharum* spp.), and fruit stones to establish orchards (Crosby, 1972:67). In the early 1500s, Spanish settlers in Mexico eagerly sought seeds and plants from Europe. In his fourth letter to Charles V of Spain in 1524, Hernán Cortés implored:

I have also informed your Caesarean Majesty of the need we have of plants of every sort, for this land is well suited to all kinds of agriculture. And because until now nothing has been sent, I once again beseech Your Majesty, as it will be a great service, to send a warrant to the Casa de Contratación in Seville that every ship shall bring a certain number of plants and shall be forbidden to sail without them, for they would be most advantageous for the colonization and prosperity of this land.[2]

The pleas of Cortés and others in Mexico paid off. The crown soon required all ships sailing to the New World to carry seeds, cuttings, roots, and livestock (McAlister, 1984:469). Starting in the sixteenth century, missionary orders also contributed to the flow of Old World plants and animals to Mexico.

[2] *Hernan Cortes: Letters from Mexico*, translated and edited by A. R. Pagden (Grossman Publishers, New York, 1971), p. 336.

With each introduction, crop plants had opportunities to adapt and change. Old World cereals, legumes, and vegetable crops brought to the New World evolved distinct types in their new setting. This process began in the late fifteenth century and continued intermittently during colonial rule. Joseph Dombey, a French plant collector and naturalist hired by Spanish and French authorities for a trip to Peru in 1777, wrote that he was bringing along "plenty of seeds and fruit stones to plant in America so as to turn over to these wild Indians with one hand what I take away from them with the other" (Steele, 1964:64).

Although the process of crop exchange accelerated during the colonial period, some genetic diversity was also lost, particularly in the New World. Aborigines of the Americas were often devastated by Old World diseases such as smallpox, tuberculosis, and the common cold introduced by explorers, traders, and missionaries; massive depopulation began in the late 1400s and continues today whenever an isolated tribe is contacted (Hemming, 1978a,b). The unleashing of Old World diseases in central Mexico, for example, was largely responsible for a 97 percent drop in the aboriginal population between 1518 and 1618 (Cook and Borah, 1979:168). Extensive tracts of cultivated lands reverted to forest or brush. Many landraces or traditional varieties of Neotropical crops undoubtedly perished along with the societies that developed them. In Hispaniola, for example, Spaniards lamented the loss of good-tasting sweet potato varieties as a result of tribal depopulation as early as 1568 (Patiño, 1963).

On balance, though, more varieties probably arose as a result of crop interchange between continents and islands during the colonial period than were lost due to cultural disintegration. Numerous landraces of cassava (*Manihot esculenta*) and maize, for example, arose in Africa after they were brought from the New World by the Portuguese (Purseglove, 1975:308). Maize reached West Africa by the second half of the sixteenth century and was widely planted on the continent except for Uganda by 1900 (Crosby, 1972:186; Cock, 1985:15). Cassava was probably brought to the Congo and Angola in the 1500s (Crosby, 1972:187). In most African villages where cassava is consumed, several local varieties are cultivated for their peculiar characteristics, including time to harvest, yield, and suitability as a vegetable or for preparing paste, flour, or starch (W. O. Jones, 1959:98).

DECLINE OF CROP GENETIC DIVERSITY

The general trend of gradual genetic enrichment of crops slowed considerably, and in some cases halted, particularly in areas where they were domesticated or they diversified, with the advent of modern plant breeding in the 1920s. As farmers adopted modern varieties and agricultural practices, they generally shifted to monoculture, and this has led to a genetic simplification of many farm lands. To feed a rapidly growing human population, plant breeders working in public and private spheres have concentrated their efforts on raising the yield ceiling of a restricted number of crops. Results have been spectacular. Traditional farmers, the early plant breeders, are as much concerned with yield stability as high productivity. A strategy of hedging one's bets by planting numerous landraces and by mixed cropping on small, unmechanized fields has prevailed for most of farming history and has resulted in generally stable but low-yielding agriculture. Scientific breeders, on the other hand, have focused more attention on developing high-yielding varieties that are fertilizer responsive as well as resistant to disease and pests. Increasingly, scientific breeders are also turning their attention to adapting varieties to difficult environments. As a result, on modern farms fewer varieties are planted in larger fields. In the United States during 1969, for example, four or fewer varieties of common bean (*Phaseolus vulgaris*), cotton (*Gossypium hirsutum*), pea (*Pisum sativum*), potato (*Solanum tuberosum*), rice, and sweet potato accounted for over half the acreage planted to each crop (Wilkes, 1983).

Until recently, much of the increase in agricultural production has come from expanding the area of tilled land, but now the trend is more towards improving productivity on existing tilled land as the earth's more hospitable regions become crowded and farming costs soar. In the past fifty years, most of the increased output from farms in the industrial nations has come from higher yields on existing farm land rather than from an expansion of the cultivated area. And the great strides made by India to feed herself in the last fifteen years are largely due to the introduction of modern varieties and improved agronomy rather than the opening up of new farm land (Plucknett and Smith, 1982; Leaf, 1983).

While yields have generally been edging upwards in this century, the genetic base of crops entering world trade has narrowed. The main forces behind the genetic erosion of the important food and cash crops in their traditional areas of cultivation, and wild relatives in their indigenous niches, include the displacement of landraces by modern varieties, a shift to monoculture, human settlements occupying the habitats of crop relatives, land clearing, the creation of reservoirs, overgrazing, fuelwood gathering, and the extinction of tribal cultures with their cornucopia of unique landraces (Timothy, 1972; Timothy and Goodman, 1979; Prescott-Allen and Prescott-Allen, 1982a; Wilkes, 1985). Urban

9

1.2. Several varieties of bread wheat (*Triticum aestivum*)
cultivated in one field in northern Syria, June 1984.

development in Mexico and Guatemala, for example, has usurped the
home of some populations of teosinte (*Zea mexicana*), the closest relative
of maize (Wilkes, 1985). In Egypt, several traditional crop varieties
were flooded when the Aswan dam closed (IBPGR, 1984a:45). Overgrazing is threatening wild relatives of most crops in the Fertile Cres-
cent in West Asia and in North Africa; even wild grasses and legumes
essential for range lands are threatened. In most arid regions, excessive
wood gathering to fuel cooking stoves threatens many perennial species
and destabilizes soils, thereby undercutting the basis of rural life. In the
industrialized world and increasingly in many parts of the Third
World, the use of machinery, consumer preferences, marketing forces,
and requirements of the food-processing industry have also prompted

the large-scale planting of a smaller pool of varieties with similar characteristics (Plucknett and Smith, 1986a). Few tomato varieties can survive shipment, for example, and the potato-chip industry has imposed limits on the number of acceptable varieties grown by commercial growers.

Species and varietal attrition is a normal evolutionary process; farmers were abandoning traditional varieties long before the arrival of scientific plant breeding. In medieval Europe, for example, people grew carrots with purple, yellow, white, and orange flesh, but the latter color type predominated well before the twentieth century. Also, some germplasm assemblages were obliterated before the advent of modern gene banks. Valuable collections of fruits, vegetables, and ornamentals acquired through introductions as far afield as West Asia, were destroyed when Henry VIII ordered the dismantling of English monasteries and their associated gardens when he severed relations with the Pope in 1534 (McClean, 1981:213). Natural catastrophes, such as widespread droughts, floods, and volcanic eruptions, have also historically wiped out landraces and populations of crop relatives.

But the tempo and scale of genetic erosion in the twentieth century are unprecedented. Field botanists and agronomists began remarking on the loss of traditional varieties in the early 1900s. J. Burtt-Davy, for example, a botanist working in South Africa, noted in 1919 that the once popular Boer oat variety was being rapidly replaced by Algerian oats.[3] Burtt-Davy collected samples of the Boer variety, formerly the predominant oat in the Transvaal where it was used for forage, and sent them to the Office of Foreign Seed and Plant Introduction, within the United States Department of Agriculture's (USDA) Bureau of Plant Industry, in Washington, D.C. In 1923, Harry V. Harlan, an agronomist with the bureau, observed that older cereal varieties in Algeria and Tunisia were being replaced by American cultivars.[4] Even in remote Saharan oases, such as Mariout in Egypt, indigenous crop types were disappearing as improved communications facilitated the introduction of new seeds.

The trend towards genetically more uniform agriculture has been especially pronounced since the 1940s (Harlan, 1975a). Since World War II, for example, virtually all of the local wheat varieties in Greece, Italy, and Cyprus have been abandoned and most of the indigenous sorghum (*Sorghum bicolor*) races of South Africa disappeared after the release of

[3] *Plant Immigrants*, No. 164, p. 1511, December 1919 (United States Department of Agriculture, Office of Foreign Seed and Plant Introduction).
[4] *Plant Immigrants*, No. 215, pp. 1969-1970, March 1924 (United States Department of Agriculture, Office of Foreign Seed and Plant Introduction).

high-yielding Texas hybrids (IBPGR, 1976). This process of more genetic uniformity of farm lands is not confined to the main cereal crops.
In the United Kingdom, many older varieties of Brussels sprouts (*Bras-
sica oleracea* var. *gemmifera*) disappeared after the release of hybrids in
the 1960s (Innes, 1975). A similar pattern prevails with many fruit trees
in Europe and North America.

As farmers see the opportunities provided by modern varieties, they
usually discard landraces, many of which subsequently vanish because
their survival depends on man. When farmers can take advantage of superior seed or clones, can obtain credit, can profitably use purchased
inputs such as fertilizers, and have easy access to sizable markets, they
frequently drop older varieties. Modern varieties generally respond
better to fertilizer than traditional varieties and may also be more resistant to pests and diseases. Even with suboptimal inputs of fertilizer
and pest control, modern cultivars often perform better than traditional varieties. The abandonment of landraces is particularly advanced
in the industrial nations and in the better agricultural zones of the developing countries, where much of the world's food and other agricultural products originate. The increasingly urbanized world depends
heavily on production from such optimal farming areas. Even in areas
where modern varieties have not made much impact, the genetic base
of food crops is often endangered. Overgrazing and rampant clearing
of pristine areas in the Third World are increasingly common, thereby
destroying wild relatives of crops with their reservoir of potentially useful genes.

DANGERS OF GENETIC SIMPLIFICATION

When large areas are planted to a single variety or a handful of cultivars
with a similar genetic background, they can be especially vulnerable to
pests, diseases, and severe weather (Baker, 1971; NAS, 1972; Harlan,
1975b; Wilkes, 1977a; Chang et al., 1979; Eckholm, 1982; Williams,
1982). Scholars and scientists have been ringing the alarm bell about
this hazardous course for nearly fifty years (Harlan and Martini, 1936;
Sauer, 1938; Hartley, 1939), but only recently has the issue been picked
up by the media.[5] Examples of the dramatic effects of genetic simplifi-

---

[5] The outbreak of southern corn leaf blight disease in the United States during 1970 triggered a series of newspaper articles on the ravages of this disease and the narrow genetic
base of maize in the Midwest and South. The value of maintaining the genetic diversity of
crops then submerged from public attention until roughly 1982. Examples of recent newspaper coverage of the dangers of genetic erosion of crops and the importance of germplasm conservation include: Nigel Smith, "Food Security and Genebanks," *The Christian*

12

cation of crops can be drawn from prehistory right up to the present. All types of crops are vulnerable to genetic simplification, ranging from food plants to industrial crops. Drastic drops in food production due to genetic simplification are the most serious, but a decline in the productivity of industrial crops is also of concern because of its economic impact, particularly with regard to jobs and purchasing power.

The collapse of the Mayan civilization, for example, may have been triggered by exclusively planting maize on extensive terraces and raised fields (Turner, 1974). Population pressure in the Yucatán peninsula prompted the construction of hillside terraces and the excavation of ridged fields in marshy areas starting around A.D. 500. The traditional pattern of slash-and-burn farming in the jungles apparently no longer provided sufficient food. The Mayans thus intensified farming by terracing slopes to check soil erosion and facilitate irrigation, and by bringing formerly untilled swamp land into year-round production. In contrast to isolated swidden plots in the forest on a variety of soil types, maize fields along terraces and raised fields were extensive and favored the buildup and spread of diseases and pests. Not only did the Mayans probably plant mostly maize on their intensively cultivated fields, they may also have planted fewer varieties of the crop in the more managed environments of terraces and ridged fields. Successive outbreaks of maize mosaic virus transmitted by corn leafhoppers (*Peregrinus maidis*) may have severely undercut maize production, possibly leading to the collapse of the Mayan civilization around A.D. 900 (Brewbaker, 1979).

Science Monitor, Boston, 16 August 1982; Norman Myers, "Billion Dollar Root," *The Guardian*, London, 23 December 1982; Dawn Weber, "Professor Studies Life Cycles of Breakfast Cereals," *Gainesville Sun*, Gainesville, Florida, 9 March 1983; "America's Favorite Potato could be Wiped out by Disease," *Pensacola Journal*, Pensacola, Florida, 16 May 1983; "Potato Facing Problems Ahead," *Lake City Reporter*, Lake City, Florida, 16 May 1983; "Potato Called Vulnerable to Pests, Disease," *The Florida Times-Union*, Jacksonville, Florida, 16 May 1983; "Future of Potatoes in U.S. 'Precarious'," *Gainesville Sun*, Gainesville, Florida, 26 May 1983; "Potatoes Becoming Extinct?" *Palatka News*, Palatka, Florida, 16 May 1983; "U.S. Potato Crop in Jeopardy?" *Daytona Beach News*, Daytona Beach, Florida, 16 May 1983; "Lack of Diversity may Spell Disaster for America's Potatoes, other Crops," *St. Petersburg Times*, St. Petersburg, Florida, 22 May 1983; "Potatoes may be in Precarious Position, UF Geographer Says," *Star Banner*, Ocala, Florida, 19 May 1983; "NCSU Experts Say Famines are Inevitable," *The News and Observer*, Raleigh, North Carolina, 25 March 1983; Hank Daniszewski, "Plant Genetic Pool Shrinking," *The Western Producer*, 24 November 1983; George Anthan, "Seed Banks Play a Critical Role in Future of U.S. Crops," *The Des Moines Register*, Des Moines, Iowa, 6 February 1984; Norman Myers, "Out of Its Skin," *The Guardian*, London, 6 June 1984; Bill Paul, "Third World Battles for Fruit of Its Seed Stocks," *Wall Street Journal*, New York, 15 June 1984; "Banco de Germoplasma de Yuca, en Palmira," *El Pueblo*, Colombia, 30 January 1984; Robert Clark, "Pour Quoi les Graines ne Meurent pas," *Le Matin*, France, 23 January 1984.

The Irish potato famine is one of the most dramatic examples of the fragility of crop productivity due to a narrow genetic base. In 1846, the late blight fungus (*Phytophthora infestans*) cut Ireland's potato production in half, provoking widespread starvation and the emigration of a quarter of the population, mainly to the eastern United States (Crist, 1971). Unlike their counterparts in the Andes, where the potato was domesticated at least eight thousand years ago and where numerous landraces are tended, Irish farmers were growing potatoes that had been multiplied from a few clones introduced from England and mainland Europe, which in turn had received material from only two samples brought from South America: to Spain in 1570 and England some twenty years later (Hawkes, 1979). Late blight disease spread quickly through Ireland's potato fields because the plants were all highly susceptible to the fungus. On the steep slopes and twisting valley floors of the Andes, in contrast, a patchwork quilt of modest-yielding landraces continues to provide some protection against the devastating disease.

A mutation of another leaf-attacking fungus, *Helminthosporium maydis*, which causes southern corn leaf blight, depressed United States corn yields by an average 15 percent in 1970 and cost farmers hundreds of millions of dollars in damage. By the late 1960s, virtually all maize production in the United States was based on hybrids developed from male-sterile lines (in which tassels [male flower] do not produce viable pollen) using a single source of cytoplasmic male sterility from Texas, a genetic trait contained within the plant cell but outside of the nucleus (Goodman, 1976). Previously, maize hybrids were produced by removing the tassels from lines to be used as the female parents, a time-consuming and expensive process (Simmonds, 1979:152). Unfortunately, hybrids containing Texas cytoplasmic male sterility were all highly susceptible to southern corn leaf blight, a disease that had lingered as a minor problem for decades until race T came along in the late 1960s (Ullstrup, 1972).

A new strain of another crop pathogen, citrus canker (*Xanthomonas campestris*), threatens the orange, grapefruit, lemon, and lime groves of Florida. Because only a handful of citrus varieties are grown in Florida, and all are highly susceptible to canker, the pathogen could easily infect the entire state and possibly spread to Texas and California. In the late summer of 1984, the causative bacterium appeared in a major nursery in the center of the state. By October of that year, 3 million seedlings had been destroyed, representing about one-fifth of the citrus nursery stock in the state (Sun, 1984a). By the end of 1984, about 6.5 million young orange trees had been destroyed in Florida in an effort to halt the spread of the disease.[6] Analysis of samples indicates that the pathogen has mutated into an especially virulent strain that threatens to undermine the state's $1.2 billion citrus business. Currently, the only effective treatment is total destruction of infected trees; in the 1920s epidemic of the disease in Florida, 20 million citrus seedlings and trees were destroyed.

In 1985, isolated cases of citrus canker were still being reported in some Florida nurseries. In August, for example, 3 million orange trees were burned in a contaminated nursery in Haines City in central Florida. More than 9 million citrus trees were destroyed in Florida between August 1984 and August 1985 due to the disease. In September 1985, the state ordered the closing of three hundred commercial nurseries in Florida for a year, as well as hundreds of nurseries operated by grove owners and companies selling ornamental plants to the public.[7]

The problem of genetic simplification of crops has surfaced in countries across the ideological spectrum. In the Soviet Union, for example, another recent case of genetic vulnerability was triggered by cold weather. Lulled by a series of relatively mild winters, farmers in the Ukraine began planting Bezostaja, a wheat variety usually grown farther south. By 1972, the popular variety was sown on 15 million hectares. Then in that year, a severe winter struck, causing a shortfall of millions of tons of winter wheat (Fischbeck, 1981:18).

Cool weather can render conditions more favorable to disease outbreaks, particularly those caused by fungi. When large areas are planted to a few varieties, weather-related disease epidemics can cause severe economic losses. In 1979, for example, blue mold (*Peronospora hyoscyami*) cost tobacco (*Nicotiana tabacum*) farmers in the eastern United States and Canada more than $240 million (Lucas, 1980). Unseasonably cool and wet weather favored the rapid spread of blue mold. The following year, the disease struck Cuba, destroying 90 percent of the cigar crop. Cigar factories were temporarily closed and 26,000 workers were laid off, thereby further hurting Cuba's economy, which is heavily dependent on the export of commodities for foreign exchange. Blue mold's blow to Cuba's economy may have contributed to President Castro's decision to permit the Mariel boat lift to the United States.

While blue mold ravaged tobacco fields in Cuba, Canada, and the United States, another crop disease outbreak triggered by cool weather was afflicting the Republic of Korea. In 1980, unusually low tempera-

[6] *Gainesville Sun*, Gainesville, Florida, 1 December 1984, p. 8A.
[7] *Gainesville Sun*, Gainesville, Florida, 18 August 1985, p. 2C; *Gainesville Sun*, Gainesville, Florida, 21 September 1985, p. 5B.

15

tures and an epidemic of rice blast disease (*Pyricularia oryzae*) forced the South Koreans to import large amounts of rice for the first time since 1977 (Chang, 1984a). The national average yield of this basic staple had increased from 3.3 to 4.9 tons per hectare following the release of high-yielding varieties in 1971. By 1979, modern varieties with similar genetic heritage were sown on three-quarters of the country's rice growing area, a precarious basis for greater productivity unless high-yielding resistant material is ready to replace susceptible cultivars.

The problem of genetic simplification has occurred with food, beverage, and other industrial crops. It has occurred in developed nations, in the Third World, and in lands governed by a wide spectrum of political ideologies. Countries with socialist economies are just as vulnerable to yield losses as a result of genetic simplification of farm lands as nations embracing free enterprise. Stalin, for example, once boasted to Winston Churchill: "We have improved the quality of the grain beyond measure. All kinds of grain used to be grown. Now no one is allowed to grow any sort but the standard Soviet grain from one end of the country to the other" (see Evans, 1975).

The erosion of plant genetic diversity is a threat wherever agriculture is experiencing rapid change and when modern breeding methods are applied to a crop. This is not to imply that scientists are to be blamed for any potential vulnerability of modern farm lands. Indeed, crop breeders have been among the most strident voices calling for action in conserving plant genetic diversity. The success of their efforts has ironically led to genetic simplification of much of modern agriculture, and they are keenly aware of this development. Some suggest that we turn the clock back and revert to traditional farming practices to increase agricultural productivity (Denevan, 1983). Traditional agricultural systems have much to teach us about ecologically sound farming methods, but to turn away from modern plant breeding as "high tech" and irrelevant to boosting agricultural productivity would not be wise. It would be difficult, for example, to arrive at a consensus about what constitutes a traditional farming system since virtually all farming areas constantly change. Also, traditional farming practices evolved in a much less urbanized and crowded world; it is difficult to envisage how such systems would support today's world population. We have thus arrived at the point of no return; modernization of agriculture and scientific plant breeding are essential if gains that have been achieved in alleviating food shortages in many regions are to be upheld. A major task now is to render modern farm lands less vulnerable to major production fluctuations.

Although the productivity of certain crops remains tenuous due to a diminished genetic foundation, the fact that no major crop failure exacerbated by genetic simplification has occurred recently on the scale of the Irish potato famine can be largely attributed to efforts to build up and use germplasm collections. Crop gene banks now serve a pivotal role in helping to maintain the productivity of farm lands in both industrial and developing nations. In the case of the southern corn leaf blight outbreak in 1970, for example, American farmers were able to buy resistant hybrids the following year because seed companies had tapped maize collections of normal cytoplasm maize in the United States, Argentina, Hungary, and Yugoslavia. Breeders employed nurseries in Hawaii, Florida, the Caribbean, and Central and South America to quickly incorporate resistance to the disease into new hybrids before planting time in spring 1971 (Ullstrup, 1972). Further examples of how plant genetic resources have been marshalled to boost and uphold crop yields are presented in Chapters 7, 8, and 9.

High-yielding agriculture rests to a large extent on an adequate pool of genetic material for breeding. In this increasingly urbanized world, the fundamental importance of sustaining food and cash crop productivity often escapes the attention of the public. City folks' links to the countryside are sporadic and are usually confined to recreation. The agricultural underpinnings of civilization are often forgotten, particularly since less than 4 percent of the population of such industrial countries as the United States and Great Britain work on farms. Concern over toxic waste disposal, air pollution, acid rain, soil erosion, and the buildup of atmospheric carbon dioxide have generally been the focus of much more public and institutional attention than the genetic diversity of crops. But the loss of genetic diversity, particularly in crop gene pools, may well be the single most serious environmental problem facing mankind (Ehrlich and Ehrlich, 1983:78).

It may not really matter that a number of "turn of the century" apple varieties in the United States or Europe are no longer grown; most were developed as sports of closely related varieties, so little genetic variability was lost. In the case of varieties developed from closely related cultivars, the use of wild material or the heterogeneous progeny from seeds provides much greater variability. Nonetheless, the loss of landraces from the crucible of primitive agriculture is infinitely more serious.

We hope to demonstrate that, although loss of genetic diversity of crops and their wild relatives is serious and is still occurring, a great deal has been accomplished in the last decade to safeguard a world heritage. Much still remains to be done in collecting plant germplasm and in improving gene banks, but the picture is much brighter now than it was in

17

the early 1970s when the issue of plant genetic resources and genetic erosion first came to the widespread attention of the scientific community and policy makers. The danger now is that with no recent drastic downturns in agricultural productivity due to the genetic simplification of crops, policy makers will assume that resources can be safely diverted to other pressing needs. A major goal of this book is to establish the importance of maintaining the momentum in germplasm conservation and upgrading collections so that agricultural yields can be further raised and sustained.

SEEDS IN DUE SEASON

Farmers need an adequate supply of robust and reliable seed in time for planting. This basic tenet of agriculture was achieved in early times quite differently than now, but the principle still holds. To highlight this theme and underscore its dependence on a well-preserved and evaluated germplasm base, we examine in this chapter the pattern of a relatively rapid turnover of varieties on modern farms. Unlike traditional farms where varieties may last for decades or generations, modern varieties usually last only a few years before they are replaced by superior material.

With regard to seed quality and sustainable agriculture, we discuss crop breeding strategies that incorporate resistance to a broad spectrum of diseases and pests. Breeding strategies for prolonging the usefulness of cultivars include incorporating more than one gene for resistance and ensuring that varieties tolerate a broad range of diseases and pests. We also explore the use and potential of varietal mixtures within the context of mobilizing genetic resources for sustainable agriculture.

International nurseries are the recruiting grounds for promising crop lines, and we emphasize their role in the production of reliable varieties. Because fewer varieties are planted at any one time in modern agriculture, productivity can be unstable unless proto-varieties are tested under a wide range of conditions. Before a farmer receives a new variety, crop lines pass through various stages, from germplasm collection to final approval by state-run agencies (Figure 2.1). Eight to fifteen years normally elapse between the time an initial cross is made in breeding programs until a crop variety is ready for release. It is thus imperative to have a number of potential varieties or proto-varieties in the pipeline at any given time to replace material that no longer holds up in farmers' fields. Therefore, finally, we analyze seed-production systems, the importance of seed certification, and the issue of plant-variety rights as it applies to crop germplasm conservation and utilization.

THE VARIETAL RELAY RACE

In modern agriculture in industrial countries and in many areas of the developing world, the traditional pattern of a mosaic of many land-

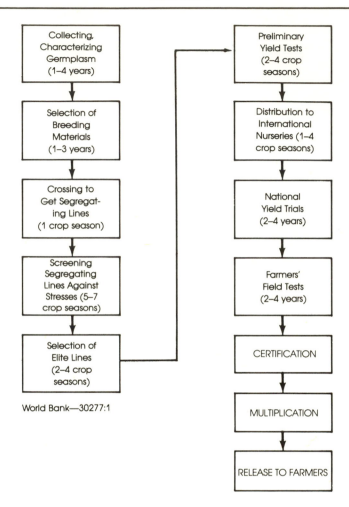

2.1. Stages and time required in the development
and testing of crop varieties.

races, all with relatively long life spans, has been replaced by fewer cultivars, each with a relatively rapid turnover (Figure 2.2). Varieties are retired when they no longer resist diseases or pests because the latter have mutated, when growing conditions or consumer preferences change, or when more promising material becomes available (Plucknett and Smith, 1986b). In developed nations, a variety of a field crop currently remains in use for only four to ten years; even when a variety still resists diseases or pests, farmers may drop it for equally resistant but better-yielding material. The high levels of modern crop yields thus depend on a steady stream of new cultivars; if this relay race faltered, crop yields would dip.

A sharp drop in agricultural productivity in the bread-basket nations of the world would trim export earnings and hurt the pocketbooks of farmers and consumers. The impact of sagging farm production in the Third World can be even more dramatic: in addition to damaging the income-earning capacity of farmers, both rural and urban poor would suffer from rising food prices; the wealthy spend a smaller portion of their income on food than the poor, making a climb in the cost of food especially painful to the latter. In developing countries with rapidly growing populations, a failure to maintain at least current crop yields could provoke widespread hunger.

The challenge of upholding gains in agricultural productivity and of further raising yields is being addressed by government agencies and private companies, with varying degrees of success, by turning to plant genetic resources and by manipulating the crop environment. This two-pronged strategy entails improving varieties and modifying adverse conditions through irrigation, fertilizers, and pest control. In optimal farm lands, both approaches are employed. But in marginal areas, irrigation may not be feasible and farmers may not be able to afford purchased inputs. In such regions, plant breeders are tailoring crops to fit difficult situations rather than relying on agronomy to mold conditions to suit the crop. This approach will become more important as population pressure increasingly drives people on to marginal lands.

Even in the more arable areas, farmers are becoming increasingly interested in crops that resist diseases and pests and take up soil nutrients more efficiently. The early success of pesticides has not always been sustained because of the development of resistant insects[1] and diseases and because of increasing restrictions on the use of agro-chemicals for environmental reasons. Over four hundred species of agricultural pests now resist one or more pesticides, and the number of insects developing resistance to applied chemicals nearly doubled between 1970 and 1980 (Sun, 1984b; May, 1985). Innate resistance to disease and pests in a crop can reduce or even eliminate the need to apply chemicals, thereby trimming farming costs. Farmers want high-yielding material but they must control their production costs. Most farms in the industrial nations are

[1] To facilitate comprehension for readers not versed in biology, we use the term "insects" to refer to all arthropods, although some arthropods, such as mites and spiders, are not classified as insects. Arthropods are joint-legged creatures; some, such as insects, have six legs; others, such as spiders, have eight legs; and others still more.

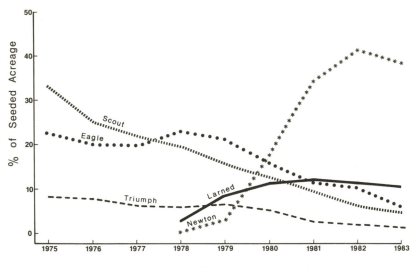

2.2. Succession of leading winter-wheat varieties
in Kansas, 1975-1983. (From KCLRS, 1983.)

business operations; owners are interested in plant types that remain productive with fewer inputs (Boyer, 1982).

Resistance breeding for insect pests and pathogens and screening plants for tolerance to environmental extremes has several advantages (Fry, 1982:195). Besides planting, no action is required by the grower, and resistant varieties do not disrupt the environment. A major bonus for farmers is that genetically resistant crops reduce operating costs. A resistant variety is sometimes sufficient by itself to suppress a disease or pest to tolerable levels. At the very least, fewer agro-chemicals are needed to control diseases and insects.

PLANT BREEDING STRATEGIES

Resistance to a single environmental challenge is rarely sufficient to maintain high yields. Even under optimal soil and climatic conditions, crops are generally attacked by a large cast of pathogens, insects, nematodes, rodents, and other pests. Thus drought resistance alone is unlikely to ensure stable yields in a crop. Yield oscillations are likely to be less severe when resistance to a broad range of disease and insect pests and tolerance to adverse weather or soils have been incorporated into a variety.

Pyramiding or concentrating genes into breeding lines that confer resistance to a broad range of environmental challenges is time consuming and difficult, but nevertheless essential to dampen yield fluctuations and maintain a steady flow of agricultural products. The rice development program of the International Rice Research Institute (IRRI), based in the Philippines, is one of the most successful examples of broad-based resistance breeding. Since the IR8 rice variety was released in 1966, numerous high-yielding varieties have been launched from IRRI's program in conjunction with national agricultural institutes and agencies. IR62 was released in the Philippines in 1984, for example. During this time span of almost two decades, the yield of semi-dwarf IRRI lines has remained essentially the same, but varieties have become progressively more resistant to a broad range of diseases and pests.

In India, the highly adaptable maize hybrid Ganga 5, developed by the Indian Agricultural Research Institute (IARI), resists brown stripe, downy mildew, leaf blight, and stemborer (Singh, 1980). Chinese wheat breeders have used germplasm from collections in several countries, including Austria, Brazil, Canada, and the United States, to develop varieties that withstand attack from a wide assortment of diseases. China harbors all major wheat diseases, making local scientists clearly skillful in averting massive crop failure due to pathogens; the last serious outbreak of a wheat disease in China occurred in 1964 in Shensi province (Johnson and Beemer, 1977).

Crop yields are also generally more stable if a variety resists more than one race of a disease or pest. Varieties resistant to several pathogen strains or pest biotypes usually have more than one gene that confers resistance, and yields are generally more durable (Van der Plank, 1968:12; Watson, 1970). When a cultivar has several genes coding for resistance or tolerance to a pest or disease, it is said to have horizontal or polygenic resistance; a crop variety with horizontal resistance survives challenges by most known races of a pathogen (Van der Plank, 1963:120).

Yield gains based on a single gene, sometimes referred to as vertical or monogenic resistance, are usually more tenuous because it is easier for a pest or pathogen to evolve a strategy for overcoming the obstacle. This principle can be envisaged if each gene is seen as a pillar holding up the yield ceiling of a crop; the more the supports, the greater the chance that yields will remain stable.

Single gene resistance can, however, sometimes hold up for several decades. In the case of the tomato (*Lycopersicon esculentum*) industry in Florida, for example, breeders at the University of Florida have been able to combat wilt (*Fusarium oxysporum lycopersici*) using a single gene; thirteen cultivars with monogenic resistance to the disease have been released in the state since 1949 (Crill et al., 1982a). Race 2 of the disease

23

appeared in 1960 but did not become widespread in Florida until the late 1960s, by which time breeders had time to develop a new variety, Walter, resistant to the newly evolved fungal strain. Walter was released in 1969 and within three years accounted for virtually all of the state's tomato acreage (Crill et al., 1982b). Typically, though, monogenic resistance breaks down within a few years. In the United States, for example, oat cultivars with the Bond gene for crown rust resistance were released to farmers in 1943, and by 1948 they accounted for 90 percent of oat acreage in the north-central United States and three-quarters of the area devoted to the crop nationwide. Then in 1949, new races of crown rust attacked the Bond-derived cultivars (Frey et al., 1973). In Iowa, average yield losses due to the new races of crown rust rose from 12 percent in 1949 to 30 percent in 1953.

Another way of improving yield stability is to plant varietal blends containing a large number of strains that share similar agronomic and marketing qualities but are based on different parents and are genetically diverse for disease resistance. This strategy is similar to that of traditional farmers who grow mixtures of plant varieties in the same field. Modern varietal blends have been used mostly with cereal crops. The Colombian national program, for example, successfully thwarted stripe rust of wheat by releasing Miramar 63, a blend of ten lines (Browning and Frey, 1969). Miramar was twice as productive as the varieties it replaced. Varietal blends have been used successfully to overcome some disease problems with Kentucky bluegrass, a popular lawn and golf-course grass in the northeastern United States. Merion emerged as one of the most popular varieties of Kentucky bluegrass because of its uniform growth and lush, dark-green appearance. Although Merion sets seed, it does so in a way that makes most of the seeds genetically identical. Merion lawns thus contain very little genetic diversity and are highly susceptible to mildew and rust diseases (Adams et al., 1971). To offset these fungal diseases, turf seedsmen are blending several bluegrass varieties to form a heterogeneous mixture that effectively resists disease epidemics while producing a fine turf.

Multilines are a more refined method of introducing genetic diversity into modern agriculture. In a multiline variety, several different sources of resistance to a particular pathogen are bred into lines that are essentially identical except for the genes that confer resistance. The proportion of seed from each line is fine-tuned each year in accordance with prevailing pathogen races (Adams et al., 1971). Multilines reduce the spread of a disease among the susceptible lines because there is less inoculum and because of the buffer effect of the resistant material (Johnson and Allen, 1975). Resistant lines trap fungal spores and retard the spread of diseases (Frey et al., 1973; Luthra and Rao, 1979). Multilines stabilize the pathogen population structure since several biotypes are maintained and no strong pressure is exerted to select for virulent new races. Multilines have been used in California to protect wheat fields against rust since the 1940s and they show promise for durum wheat cultivation in India (Suneson, 1960; Pandey, 1984).

Multiline cultivation avoids the boom-and-bust cycle evident in some cases of monocropping (planting one crop per field) with modern varieties. Pests and pathogens continue to evolve when multilines are planted, but the host, pathogens, and pests are in approximate equilibrium. And with multilines and varietal blends, highly mechanized farmers can plant genetically different cultivars and still avoid such problems as staggered ripening times, large differences in plant height, or fruit with different packing characteristics.

Although multilines and varietal blends increase the genetic diversity of commercial monocrops, what makes ecological sense does not necessarily translate into farming practice. Multilines and varietal blends have found a niche in agriculture and show promise of further use, particularly in areas where fungal diseases of cereals are a major problem, as in the case of powdery mildew and barley in the United Kingdom (Wolf and Barrett, 1980). They also have a place with perennial crops, such as fuel-wood plantations, since susceptible material is more costly to replace than in annuals, which can be replaced more frequently. But they have not yet been widely adopted. One reason for their slow takeoff is that they are relatively expensive and time consuming to develop (Browning and Frey, 1969; Crill et al., 1982c). Breeders working on varietal blends must come up with several "winning" lines in terms of yield, agronomic characteristics, and market preference at the same time. In the case of wheat, for example, a buffer of 45 centimeters is recommended between lines susceptible to the same race of a disease; some fifteen different genotypes are required to make an adequate varietal blend or multiline (Watson, 1979).

Some breeders have resisted varietal blends and multilines because they perceive them as a shotgun approach to tackling disease problems. Virtually the entire arsenal of resistance genes is simultaneously brought out without leaving any in reserve for future use (Van der Plank, 1968:142). Some breeders are nervous that by exposing all known resistance genes to a pathogen, a highly virulent "super" race will evolve and undercut crop production. Four years of testing wheat multilines at nine locations in India, however, showed that the multilines continue to be more resistant than pure line varieties (Gill et al.,

25

1984). Although the wheat multilines showed some disease symptoms, damage was minimal.

Replacement of varieties in relay-race fashion is the predominant pattern in modern agriculture and will probably remain so for the foreseeable future. Multilines and varietal blends, in spite of their costly development, nevertheless have a role to play in stabilizing yields, and their use is likely to increase. Whether a conveyer-belt system of producing new varieties is used, or multilines and varietal blends are deployed, all three strategies for raising and upholding yields rely on extensive germplasm collections that have been well preserved and evaluated.

INTERNATIONAL NURSERIES

International nurseries are specialized experimental plots for advanced breeding lines and serve as launching pads for crop varieties (Plucknett and Smith, 1984). Nurseries covering many different environments across the world are the recruiting and testing grounds for promising crop lines. Whether on the steamy plains of the Punjab or a dusty valley in the Sudan, entries run a gauntlet of tests to help breeders spot desirable material. International nurseries have become indispensable to the screening of elite breeding material for wide adaptability as well as for resistance to specific disease and pest pressures.

International nurseries can be separated into two main categories— those set up to test for wide adaptability and yield, and nurseries in "hot spot" locations where germplasm is checked for resistance to specific diseases or insect pests or for tolerance to adverse environmental conditions. Nurseries screening for wide adaptability often contain thousands of entries from dozens of countries (Figure 2.3). The IRRI-coordinated International Rice Testing Program embraces eight hundred scientists in seventy-five countries in Asia, Africa, Latin America, Oceania, and Europe. International nurseries coordinated by the International Wheat and Maize Improvement Center (CIMMYT—Centro Internacional de Mejoramiento de Maiz y Trigo) also attract a large number of participants. In 1983, for example, scientists in ninety-one countries requested 2,072 trials of wheat, triticale, and barley from the Mexico-based center (CIMMYT, 1984:18). Countries participating in international nurseries try out their own materials and have the opportunity to observe the performance of foreign material that may be suitable for their conditions. Breeders looking for wide adaptability get a reading on how well breeding lines hold up across a broad range of environments.

Specialized nurseries usually have fewer participants and screen germplasm for resistance or tolerance to a specific disease, pest, soil, or climatic condition. IRRI coordinates eleven such nurseries and the CIMMYT wheat program operates several disease-screening nurseries. CIMMYT's Regional Disease and Insect Screening Nursery (RDISN) is administered from Cairo, Egypt, in cooperation with the International Center for Agricultural Research in the Dry Areas (ICARDA), while the Latin American Disease and Insect Screening Nursery (VEOLA) is administered from Quito, Ecuador (Dublin and Rajaram, 1982). Entries sown in the RDISN and VEOLA nurseries are obtained from national programs in their respective regions. Specialized nurseries designed to screen for disease resistance often contain commercial cultivars in addition to elite breeding material and provide early warning in the event a new pathogen race emerges.

International nurseries are springboards for the development of many crop varieties. CIMMYT's maize program and collaborating national programs, for example, have used international nurseries to identify promising maize material that has led to the release of over 150 varieties and hybrids which now grow on some 5 million hectares in thirty-nine countries (Sprague and Paliwal, 1984). Cereals dominate international nurseries, but germplasm of other crops is also tested over a wide range of conditions. The International Bean Yield and Adaptation Nursery, coordinated by CIAT (Centro Internacional de Agricultura Tropical) near Cali, Colombia, for example, is responsible for the launching of over fifty varieties.

Many international nurseries are coordinated by international agricultural research centers within the Consultative Group on International Agricultural Research (CGIAR). The CGIAR, headquartered at the World Bank in Washington, D.C., was established in 1971 to support research on food crops and certain livestock problems of the Third World (Plucknett and Smith, 1982). Thirteen centers, including the International Board for Plant Genetic Resources (IBPGR), now operate in Latin America, Africa, the Middle East, and Asia with the support of forty donors, ranging from national governments and multilateral and bilateral agencies to private foundations (Figure 2.4; Appendix 1). CGIAR centers also maintain many of the larger crop germplasm collections that are heavily used by breeding programs worldwide.

SEED PRODUCTION AND CERTIFICATION

Farmers traditionally either select seed from the best plants or save a small random sample from the harvested grain for the next planting

2.3. Harvesting a trial of chickpea (*Cicer arietinum*) at the International Center for Agricultural Research in the Dry Areas near Aleppo, Syria, May 1984.

cycle. Whether seed is selected randomly or deliberately, crops normally harvested at immature stages, such as runner bean, have to be grown out to maturity to gather the seed. The saving of seed from one harvest to the next is still widespread among farmers planting landraces, particularly in marginal areas. Farmers' storage bins and markets can be rich, though temporary, accumulations of germplasm.

Farmers always keep a keen eye open for particularly vigorous or unusual plants, and progeny from such plants are often carefully set aside. The saving of seed from superior plant types was a forerunner to modern plant breeding in which superior lines are selected from variation created by controlled crosses using selected parents.

Sometimes a farmer has surplus seed, and informal exchanges between local farmers are still common in some regions. Such exchanges, often involving bartering rather than cash settlements, may result in a "best local" variety becoming fairly widely adopted within a region. This pattern still prevails in much of the Third World, particularly in remote areas. And in industrial nations, some gardeners participate in unstructured marketing systems for seed sales, as through catalogs or at fairs.

In this century, organized seed companies have arisen to meet the need for large quantities of top-quality seed of better-yielding varieties. Organized seed production and seed-technology research began in ear-

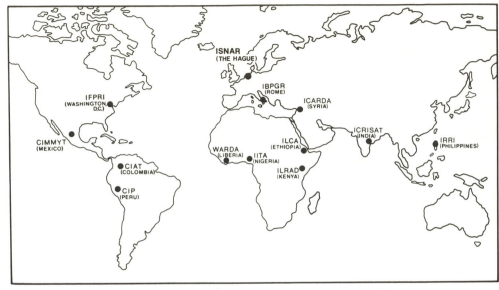

2.4. International agricultural research centers within the Consultative Group on International Agricultural Research.

nest in the early 1900s, although their origins can be found earlier in Scandinavia and the United States. In the United States, for example, the Hatch Act was enacted in 1875 to promote the seed industry, although seed certification came much later. As with any commercial activity, distrust relating to the quality of the product and advertising sometimes arises, but on the whole legitimate seed companies have established themselves both in industrial nations and developing countries to supply products to increasingly large numbers of market-oriented farmers.

Seed companies have generally prospered because they respond to three main concerns of farmers. First, farmers want options when seeking high-yielding varieties to plant. Second, farmers must have reliable seed that performs well for them. A seed company will not stay in business long if it does not produce a trustworthy product. In most cases, a government agency oversees this quality-guarantee stage by certifying seed after verifying its suitability. Third, numerous seed companies, often small, have arisen to supply products to growers specializing in relatively minor crops.

New varieties must reach the farmer before current cultivars fail due to changing pests or other environmental stresses, and seed must arrive on the farm at the proper time for planting. A supply of high-quality seed is especially critical to modern farmers who rely on fewer varieties than traditional farmers.

A large seed industry has developed in areas often remote from where the seed is planted, and seeds may be multiplied and marketed over wide areas. Pioneer Hi-Bred, for example, a maize seed company based in Des Moines, Iowa, operates in dozens of countries and annually sells more than half a billion dollars worth of seeds (Kahn, 1985:68). Public and private seed operations in the industrial nations take advantage of dry weather during the harvest season to produce seed and often increase seed at a number of widely separated locations. In some cases, tropical and subtropical stations are used to generate two seed crops per year. Field locations in Hawaii, for example, have been used for this purpose for almost two decades.

Many generations of seed selection and production away from the area in which the crop is to be grown can lead to genetic changes and a poorly adapted variety. Independent seed certification agencies safeguard the farmer against such potential losses. In cases where the breeding strategy focuses on wide adaptability rather than on tailoring varieties to specific environments, this problem is less acute. Many of the international agricultural research centers are breeding for wide adaptability so that material is appropriate for planting over large areas, or can be readily adapted with some further crossing by national programs.

In some developing countries, the availability of good-quality seed of improved varieties has lagged behind the spectacular gains made through plant breeding over the past three or four decades. In some cases, particularly in smaller and poorer countries, the vital link between the development of high-yielding varieties and the multiplication of sufficient good-quality seed was never made, or is broken. To try and close this gap, the Food and Agriculture Organization (FAO) launched a World Seed Campaign in 1957 in which seventy-nine countries participated. At the culmination of the campaign in 1961, governments were more aware of the need for mass production, multiplication, and certification of good-quality seed.

Major constraints in developing good seed-production systems still exist, but progress is being made. Over four hundred seed production projects have recently been implemented in eighty countries, and over 2,500 seed specialists have been trained (FAO, 1985). The Washington-based International Agricultural Development Service (IADS) is helping the Nepalese Agriculture Inputs Corporation establish a labor-intensive system to produce, test, process, store, and distribute seeds of

31

the major food crops, especially in the hills where transportation is difficult. With the support of FAO and IADS, Turkey has developed a project to produce, process, and market synthetic maize varieties, forages, and vegetables. Several international agricultural research centers within the CGIAR system have become heavily involved in seed production; the seed unit at CIAT has been so successful that its coordinator, J. E. Douglas (1980), has produced a planning and management guide that is widely used throughout the world.

Before seeds reach the hands of farmers, they are usually certified by an independent government agency. Seed-certification agencies check seeds or clones for purity, quality, and health. In checking for purity, technicians search for seeds of weeds and stray varieties. Seed-producing operations try to ensure purity by good weed control and keeping seed-production fields away from areas planted for direct consumption. Quality tests check for viability, and seeds are inspected to reduce the chances of spreading diseases and pests. In the case of clones, purity is rarely a problem, but it is harder to check for quality and health. The certification process normally delays the release of a variety for up to three years.

With regard to standards for seed certification, the International Seed Testing Association (ISTA) promulgates procedures and methods for testing, and recognizes seed-testing laboratories throughout the world. When countries become more efficient in seed-quality testing and certification, ISTA provides counsel on appropriate seed legislation. Most developing countries also seek advice from other nations with a long history of seed certification when they are considering such legislation.

In addition to ISTA, other regional organizations establish seed-certification standards. The Organization for Economic Cooperation and Development (OECD) sets standards for seed certification for member countries. The OECD recognizes the following steps in seed production: (1) pre-basic, (2) basic (both steps produced by the breeder), and (3) successful certified generations. In North America, the Association of Official Seed Certifying Agencies (AOSCA) follows similar steps (Feistritzer, 1975).

In spite of seed-quality control measures, some concern has been expressed that multinational corporations exercise a tight grip on the supply of seed to developing countries and are not acting in the best interest of small farmers. Such allegations fail to take into account the beneficial impact of multinational seed companies. For instance, twenty companies donated 37,997 samples of vegetables to seventeen African countries to help FAO's program to rehabilitate food production on the continent (FAO, 1985). Many of the donated samples were distributed to small farmers, women's cooperatives, and refugee camps.

Some multinational seed corporations are owned by oil and pharmaceutical companies, and critics of multinational seed companies contend that they promote expensive packages of herbicides, pesticides, fertilizers, and fungicides along with seeds (Mooney, 1979:86). While it is true that some multinational seed companies are owned by oil and pharmaceutical corporations, no seed company stays in business long if its products are not suitable to farmers. The issue is not whether capital inputs are necessary in modern agriculture; the question is whether they are cost effective. Farmers are not interested in the highest possible yield regardless of input costs. Farmers only apply fertilizers and other chemical inputs if they boost yields enough to at least cover their costs. Besides, private seed companies offer varieties with varying degrees of resistance to pests and diseases; farmers need not become chronically dependent on unnecessary inputs.

The problem is not so much that multinational seed companies will dominate the agricultural scene in developing nations; the issue is getting them to invest in Third World countries. In some cases, seed companies have explored the possibility of setting up operations in a developing country only to be told that they cannot repatriate any profits. Some Third World countries prohibit the establishment of private seed companies, while other nations are not attractive to multinational seed companies because the market is too small, areas with good soils and an assured water supply are limited, and infrastructure is poor. In 1984, for example, U.S. seed companies sold only $328 million worth of seeds overseas, less than 2 percent of their total seed sales (Witt, 1985:97). Where private seed companies have been allowed to operate, agricultural productivity is generally higher than in countries where they are outlawed. Mexico, Brazil, Zimbabwe, and Kenya, for example, allow private seed companies to operate and these countries are known for their dynamic agriculture.

Burdened by mounting food import bills, developing countries are becoming increasingly interested in hybrid crops. Sri Lanka, for example, is interested in hybrid potatoes using true or botanical seed (CIP, 1985:25). Potato production based on botanical seed, as opposed to seed potatoes, frees the farmer from having to set aside about 12 percent of the harvest for the next planting. Interest in hybrid crops is particularly keen with regard to cereals. In Nigeria, two private seed companies were established in 1984, and a third began operating in 1985, to produce and market hybrid maize seed (IITA, 1985:58).

Hybrids usually produce between 10 and 30 percent more grain than open pollinated varieties and consumers benefit from the increased production through lower prices (Griliches, 1958). Hybrids can be produced by crossing inbred lines or an inbred line with a high-yielding variety (topcross); the resulting increased growth is termed heterosis or hybrid vigor. Harvested seed from hybrids cannot, however, be successfully used for the next planting because the progeny segregate into a mixture containing runts, mediocre plants, and only a few superior specimens. Thus a farmer must buy fresh seed each year to ensure continued high yields.

Despite this constraint, hybrids have revolutionized a growing number of agricultural regions, such as the Corn Belt in the United States, maize-growing areas of Kenya and Zimbabwe, some sorghum-growing parts of India, and rice lands in mainland China. In the 1950s, FAO promoted the transfer of maize hybrids to southern and eastern Europe and the Mediterranean, resulting in an average 80 percent increase in yields within a decade, from 1,240 kilograms per hectare in 1953 to 2,040 kilograms per hectare in 1962. In Hungary, the maize area devoted to hybrids climbed from 3 percent in 1957 to 100 percent in 1964 (Kahn, 1985:70). The Peoples' Republic of China is the only country with sizable plantings of hybrid rice. The Chinese pioneered hybrid rice research, and production is firmly in the hands of government agencies. Hybrid rices now cover one-quarter of the country's vast rice area. And in the Indian states of Maharashtra and Karnataka, hybrid sorghums occupy two-thirds of the sorghum-growing area (Hawkes, 1985:33).

Although the success of hybrids in developed countries and parts of the Third World is not in dispute, private seed companies and the use of hybrids are still controversial in developing countries. Some argue that poor farmers in developing countries have a right to such technology, whereas others fear that farmers will become too dependent on expensive inputs and will be exploited. Considering the sharply opposing views on the subject, it is worth briefly to review the history and performance of the two independent seed companies in Kenya and Zimbabwe.

The Kenya Seed Company was founded in Kitale in 1956 to foster the use of improved pasture seed, but soon expanded operations to cover seed production of hybrid maize, barley, wheat, sunflower, and horticultural crops. The company is best known for its high-quality hybrid maize seed, which it began to produce in 1963. From 1963 to 1973, over 80 percent of farmers in the major maize-growing region of western Kenya were planting hybrids, an adoption rate faster than in the United States Corn Belt in the 1920s and 1930s.

Hybrid seed production is contracted out to individual farmers and is checked for quality by company staff as well as the government-operated Inspection Service for Seed. Certified seed produced by the Kenya Seed Company is distributed to farmers through a network of over six thousand small shopkeepers selected for their location and reputation. The company has established two or three distributors for each village to avoid a local monopoly and to promote competition. Hybrid maize seed produced by the Kenya Seed Company is sought after by neighboring countries, and after satisfying internal demands, principally from small farmers, seed is exported. In 1980, the company exported 3,000 tons of hybrid maize seed, mainly to Uganda, Tanzania, Ethiopia, and Sudan. The company has thirteen extension staff, five of whom work with maize, to advise farmers and receive feedback on problems.

The Zimbabwe Seed Company, still in private hands, has produced a number of successful maize hybrids. One hybrid, SR52, is particularly popular and has been adopted in neighboring Zambia. Zimbabwe's agriculture is relatively vigorous and, with the return of near-normal rains in 1985, the harvest will be double the country's internal requirements (Brown and Wolf, 1985:39). Zimbabwe is one of the few countries in Africa that exports food.

The success of the Kenya Seed Company and the Zimbabwe Seed Company can be attributed to several factors. First, both governments give high priority to agricultural development. Second, national research programs backstop the seed companies with breeding material. In Kenya, for example, early work on improving maize germplasm is performed by the Kenyan Maize Research Program. Third, government interference in company operations is minimal. In the case of the Kenya Seed Company, 51 percent of the company stock was held by a parastatal organization by 1980, but the company still runs as a private concern. And the socialist agenda of the Mugabe government has not enmeshed the private Zimbabwe Seed Company. Fourth, the seed companies cater to farmers' wishes, not to the dictates of centralized bureaucracies. In the case of the Kenya Seed Company, for example, only white maize is marketed and farmers are offered a choice of hybrids. The company knows that farmers will forego a modest yield advantage in favor of a variety with a preferred grain type. Finally, the companies maintain high-quality standards in seed production by constantly monitoring farmers involved in seed multiplication.

We do not wish to imply that state-run seed companies are incapable of producing the quality seed that farmers demand. The Peoples' Republic of China, India, and some Eastern European countries, such as

Yugoslavia and Hungary, have state-owned seed production schemes that respond to farmers' needs and are scientifically sound (Jain and Banerjee, 1982). But government-run seed operations are generally not as accountable as private companies; it is doubtful whether a government employee has ever lost his job because he delivered poor-quality seed or because farmers did not receive seed in time for planting.

Quality control is especially difficult in developing countries with inadequate government supervision and qualified personnel. In a pilot hybrid maize-production scheme recently tried by one West African government, for example, some of the growers contracted to produce hybrid seed purchased their seed in markets in order to meet their quotas.

PLANT VARIETY RIGHTS

Of all the issues relating to crop genetic resources, plant variety rights (also known as plant breeders' rights) is one of the most contentious. Champions of plant breeders' rights argue that only when patents or royalties can be applied to the finished products of research is there sufficient stimulus for research and development in agriculture, or any other scientific enterprise. Others argue that genes are a world heritage and in the public domain and cannot be patented. Mooney (1983), for example, asserts that private seed companies should not be allowed to exploit the Third World by employing germplasm from developing countries and selling it back to them. Furthermore, critics of plant breeders' rights suggest that patents will trigger greater secrecy in agricultural research and may hinder the flow of germplasm (Jain, 1982). Fears are also expressed that crop genetic diversity will be further reduced because of the interest of private breeders in crop uniformity.

Crop germplasm has been shifted between regions for millennia, and the pedigree of a modern crop variety usually leads back to many countries. The amount of germplasm being tested and exchanged today is impressive and illustrates what is possible when governments and scientists work together on a common mission. Unimproved germplasm, such as landraces and wild material as well as obsolete cultivars, has been, and should always be, freely exchanged. This policy of free availability for relatively unimproved germplasm is true for most crops, particularly the staple food crops.

Nationally supported agricultural research programs as well as international centers are heavily involved in freely sharing breeding material and making it available to public and private breeding programs. The international agricultural research centers are among the biggest

36

breeding operations in the world, and their products are free. National research programs, international centers, and others can use any of the finished products from private plant-breeding concerns for further breeding without any hindrance.

Advanced breeding lines, such as hybrid parent lines or mutants, as well as finished varieties, are not always freely available nor should one expect them to be so. Nevertheless, the original materials from which they were derived are available. If a company has invested considerable financial resources in developing an advanced breeding line, its desire to recoup investments and a reasonable return after years of work is understandable. If companies are denied the right to obtain royalties or to market their commercial cultivars on an exclusive basis, then the incentive for private plant-breeding efforts disappears.

Some see no role for private concerns in agricultural development in the Third World. But as Lester Brown points out in an incisive editorial in *Science* (12 September 1980), Karl Marx was a city boy and agriculture rarely thrives under his dictums. The much-maligned profit motive is a lasting and valuable stimulus to development. This is not the place to embark on an extended discussion of the merits and pitfalls of capitalist and socialist approaches to development; our point is that the public and private sectors have complementary, not mutually exclusive, roles to play in boosting agricultural productivity, particularly in conserving and using plant germplasm.

The claim that private companies exploit Third World germplasm at the expense of developing countries has been muddied by ideology and misinformation. Private seed companies dealing with food crops, it is argued, are almost exclusively concerned with hybrids because farmers must return to the company to obtain quality seed every year. In truth, maize, pearl millet (*Pennisetum typhoides*), and lately rice in China are the only major food crops for which hybrids have been developed on a large scale. In developed countries, hybrid sorghum for industrial use or animal feed has also become important.

Much of the concern about plant variety rights and the related issue of monopolization of seed supplies by multinational companies is confusing because of a misunderstanding of the process of genetic erosion and what constitutes genetic resources. Gene banks were not established to conserve breeders' lines, although elite breeding materials are sometimes kept in medium-term collections. Gene banks were mainly established to hold relatively primitive materials containing all the genes that breeders need for research and experimentation. Germplasm collections are used by breeders at international, regional, and national centers to produce finished varieties, with no attached plant breeders' rights. Crop gene banks are also used by breeders working for private companies; finished varieties stemming from their research are usually protected by legislation, but the product rarely remains popular for long.

More than twenty countries have enacted plant variety rights. Although more studies of the impact of plant breeders' rights in industrial nations are warranted, no evidence has emerged that this trend is stifling the flow of unimproved germplasm. On the contrary, the lack of plant variety protection may hinder germplasm movement because private companies cannot prosecute other companies that take their finished product and multiply it for sale. Breeders' varieties, which are subject to plant breeders' rights in some countries, are sought after by other breeders for parents in further crossing work. In countries that have enacted plant breeders' rights, breeders need not hold on to their varieties because they are protected. But where there is no such protection, exchange of varieties would depend on personal assurances that the material would not be sold directly. Such promises can be hard to keep since there is no guarantee that the varieties will not be misappropriated by others (Frankel, 1981).

Each country must weigh the benefits and possible drawbacks of enacting legislation that protects crop varieties in the context of its own peculiar history and socio-economic conditions. But one of the major benefits of plant breeders' rights is that they hasten development of seed supply systems. Those countries that deny plant breeders' rights, or procrastinate in coming to a decision on the issue, may find that they do not have access to certain high-yielding varieties produced by private companies (Godden, 1984). Indecisiveness may also be costly.

In the United States, patents have been awarded for agricultural inventions, mostly in the engineering and chemical fields, since the Patent Act of 1790 (Evenson, 1983). It was not until the Plant Patent Act of 1930 that clonally-reproduced plants, mostly orchard trees and ornamentals, were provided some protection. Sexually reproduced crop varieties achieved protection in 1970 with the passage of the Plant Variety Protection Act (amended in 1980). The 1970 act, administered by the U.S. Department of Agriculture, protects the breeder of a new, stable, and uniform variety from a seedsman who might want to reproduce that variety for sale. Farmers are free to reproduce seed from the variety for their own use and may even sell limited amounts of seed. No restrictions apply on the use of protected varieties for research.

Plant variety rights in the United States and most other countries with advanced economies have stimulated private investment in crop-breeding research (Duvick, 1983). As of 1983, certificates or patents have been granted for eighty crops in the United States, and most certificates have been awarded to the private sector (Evenson, 1983). Soybean (*Glycine max*) breeding in the United States illustrates the stimulus provided by plant variety rights. Prior to 1970, only a dozen breeding programs were operating in the public and private sectors. After the enactment of the Plant Variety Protection Act, thirty-five additional private soybean-breeding programs were launched. Of the 281 certificates for soybean varieties awarded between 1971 and 1983, 87 percent went to thirty-five independent private firms (Evenson, 1983). Over half the soybean acreage in the United States is expected to be planted to private varieties within a few years. And a 1985 decision of an Appeals Board of the U.S. Patent and Trademark Office allows genetically engineered plants, seeds, and tissue cultures to be patented; this step will be a boon for the biotechnology industry (Beardsley, 1985; Sun, 1985).

Plant variety rights do not mean the demise of public institutions' involvement in plant breeding. Over 80 percent of the varieties of wheat, rye, oats, barley, peanut or groundnut (*Arachis hypogaea*), dry edible beans, and forage grasses planted by United States farmers are public varieties (Pioneer, 1982). And many of the smaller seed companies, for example, continue to depend on the U.S. Department of Agriculture, the State Agricultural Experiment Stations, and universities for breeding material.

The advent of plant variety rights has prompted public agricultural-research institutions in the United States to adjust some of their programs. With some crops, such as tomatoes, peas, lettuce, tobacco, and onions, the public sector has virtually withdrawn from varietal development. As the private sector has grown and matured in the development of soybean and hybrid maize and sorghum, public agencies have channeled more research effort into breeding methodology, improvement of breeding stocks, and genetics. Public agricultural-research institutions, particularly in countries with advanced economies, are tending to release segregating progeny (material that has not yet achieved pure line status) of crossing programs at earlier stages to increasingly experienced breeders in national programs. Also, they are directing more of their future efforts toward basic research.

No evidence has emerged that protection of private plant variety rights impedes the conservation, exchange, and utilization of crop germplasm. Plant variety rights encourage innovation in plant breeding by providing a financial incentive. They provide for the use of patented material in research and limit the use of the variety only in countries with plant-protection laws. In nations not recognizing plant breeders' rights, patented varieties can be used without legal problems.

In most cases, private seed companies obtain patents for their products in major markets; they are not especially concerned about their free use elsewhere.

Germplasm collections in private hands are small; they rarely, if ever, hold diversity that is not available elsewhere, and they are mostly confined to advanced breeding lines. Basic or unimproved germplasm is freely available in numerous gene banks operated by national, regional, or international institutions. Gene banks are not subservient to the interests of private seed companies; rather, they backstop breeding programs operated by public institutions and companies.

PLANT COLLECTORS AND GENE BANKS

For most of Earth's history the genetic diversity of plants has been preserved in wild habitats or in farmers' fields. Besides the overgrazing by domestic cattle and goats in parts of the Middle East and Central America, nothing has significantly threatened wild relatives of our crop plants until this century. The genetic richness of crops has been maintained for at least 10,000 years by farmers saving seed or clones for the next planting, usually within a year. Seeds for the next sowing have been protected from insects and mammals by burying them in baskets surrounded by ash, sealing them in domed adobe structures, or packing them in elevated thatched huts. Vegetatively propagated crops have been historically maintained by storing tubers for a few months in cool, dry locations, such as potatoes (*Solanum* spp.) in the Andes; stacking stem cuttings or stakes away from moisture for short periods, as in the case of cassava; or by immediately planting fresh cuttings from plants, as with the sweet potato. Genetic stocks were thus maintained in farmers' plots and dooryard gardens, while crop relatives flourished in waste places and natural habitats.

In this chapter we explore the historical roots of germplasm preservation. We examine the origins of botanic gardens and analyze their functions, emphasizing crops rather than ornamental plants. Although some prefer to define botanic gardens as institutions with scientific staff and herbaria that essentially began in the sixteenth century, we include pleasure gardens, classic botanic gardens, and nurseries that span millennia. After discussing the role of botanic gardens in acclimatizing and disseminating exotic germplasm of economic plants, we turn to the rise of modern gene banks. Finally, we examine the lives and accomplishments of plant explorers without whose efforts germplasm collections would not have become possible.

BOTANIC GARDENS

The practice of assembling plants from foreign lands and keeping them in botanic gardens is ancient, and religion and mythology have often motivated their construction. The Kayapó Indians of the Brazilian Amazon, for example, maintain germplasm collections in hillside gardens

(Kerr and Posey, 1984; Posey, 1984, 1985). Kayapó gene banks, containing mostly tuberous plants belonging to such families as the Zingiberaceae, Araceae, and Marantaceae, are designed to safeguard the genetic diversity of crops in the event of a disaster, particularly from floods. A mythological concern for a massive flood is one reason why the Kayapó take the trouble to set aside representative samples of some of their important crops on high ground (Posey, pers. comm.). These specialized gardens are tended exclusively by elder women under the direction of the Kayapó female chief, the highest rank that can be attained by women in Kayapó society.

Kayapó germplasm collections are assembled in two ways. The underbrush in 8 to 10-year-old fields is cleared and tuber stocks are planted; at harvest, representatives of the varieties are left behind to preserve the collection. In new gardens, some two dozen varieties of edible tubers and numerous medicinals are planted in the shaded micro-environment of banana patches. When the field is eventually abandoned, banana suckers are transplanted in new fields and the companion germplasm collection is transferred to a mature clump of bananas. This careful attention to preserving the genetic diversity of cultivated plants has probably been practiced by the Kayapó for a long time.

In the Middle Ages, pleasure gardens of Europe and the Middle East were often conceived as a simulation of the Garden of Eden; the caring of ornamentals and fruit trees was thus regarded as reliving the creation (Glacken, 1976:347). In the Old Testament, the Garden of Eden is depicted as the first home of man and was planted by God with trees continually bearing flowers and fruit.[1] When the Garden of Eden was not discovered by navigators and overland travelers, it was thought to have perished in the great flood, and the idea emerged of bringing together the scattered pieces of the creation in a garden (Prest, 1981:9). Monastic gardens were common in Europe by the ninth century.

A combination of pleasure and spiritual sentiment also spurred the creation of gardens around mosques and Islamic palaces during the Middle Ages. While the Iberian peninsula was ruled by Moors in the tenth and eleventh centuries, for example, several impressive gardens were planted around palaces. The immense garden laid out at Medina Azahara near Córdoba under instructions from Moslem caliphs con-

[1] The King James version of Genesis 2, verses 8-10, reads: "And the Lord God planted a garden eastward in Eden; and there he put the man whom he had formed. And out of the ground made the Lord God to grow every tree that is pleasant to the sight, and good for food; the tree of life also in the midst of the garden, and the tree of knowledge of good and evil. And a river went out of Eden to water the garden; and from thence it was parted, and became into four heads."

tained irrigated ornamentals and orchards. Cool gardens provided pleasant opportunities to relax and contemplate the qualities of Allah. According to al-Maqqari, the palace gardens at Murcia were "filled with scented flowers, singing birds, and water-wheels with rumourous sound" (Harvey, 1981:43). Similar pleasure gardens were laid out around palaces in Toledo and Seville. Caliphs and sultans spared no expense in enriching these gardens; collectors were dispatched as far away as India to gather rare plants and seeds.

Gardens established by the Moors in Spain served for pleasure, inspiration, and study. The performance of introduced plants was scrutinized and experiments were made. Islamic botanic gardens were instrumental in furthering an early understanding of plant properties. Ibn al-Baitar, a leading Moorish botanist who lived in Malaga, described 1,400 plants in his pharmacopoeia (Harvey, 1981:43).

The spread of Islam has also influenced Asian gardens, where the paradise theme is again evident, as is the theme of water as a life-giving element. Moghul rulers of India, for example, established irrigated gardens in their palace grounds and around certain tombs during the sixteenth and seventeenth centuries (Crowe et al., 1972:42). Moghul gardens contained large collections of exotic fruit trees and ornamentals and were used for repose, contemplation, and family gatherings. The Moghul gardens at Agra, India, contained flowers as well as apple, orange, mulberry, mango (*Mangifera indica*), coconut, fig (*Ficus carica*), and banana trees (Crowe et al., 1972:76). The Moghuls brought cherries and apricots to India and their expansive gardens rank among the great landscape traditions of the world.

Medicine also provided an early impetus for the collection and scientific study of plants. One of the world's earliest gardens was a collection of Chinese medicinals instigated by the Emperor Sheng Nong in 2,800 B.C. (Sheng-ji, 1984:7). In the Son dynasty (A.D. 420-479), all the plants in the medicinal garden known as Du-Lee Garden had labels. A Jewish physician called Solomon maintained England's earliest private herb garden at Norwich in 1266. Solomon obtained medicinal plants from the continent (Harvey, 1981:78). Venice had a physic garden in 1333, and Pope Nicholas V ordered a plot of medicinal plants established at the Vatican in 1447 (Hyams and MacQuitty, 1969:16). Pisa and Padua had formal gardens of medicinals used for instruction in 1543 and 1545, respectively. The origins of the Leyden Botanic Garden (established in 1587), the Paris Jardin des Plantes (1626), and the botanical gardens of Oxford (1621), Edinburgh (1670), and Leningrad (1714) can also be traced to the need to accumulate and study medicinal plants (N. Smith, 1986).

Unlike most of its earlier counterparts, the world's best-known botanic garden, the Royal Botanic Gardens at Kew, London, was not established to propagate only medicinal plants. This was probably due to its relatively late start in 1759; the first official director of the Kew gardens was not appointed until 1841 (Hepper, 1982:32). For two centuries, the staff at Kew have assembled a diverse range of plants, now numbering some fifty thousand species, from virtually every country. Collectors continue to be dispatched all over the world in search of potentially useful ornamentals, medicinal plants, and wild relatives of crops.

Colonial powers established numerous botanic gardens in their overseas possessions, particularly in the West Indies and Southeast Asia (Figure 3.1).[2] The French were responsible for the first botanic garden in the tropics, which was built in 1735 at Pamplemousses on Mauritius (MacPhail, 1972:102). Pamplemousses, as with most botanic gardens established by colonial powers, emphasized economic plants. At its inception, it was planted with vegetables, cassava, and pasture to provide food for the French community. Fruit trees were added within a couple of years and Pamplemousses, the French word for grapefruits, took on the additional role of a pleasure garden (Hart, 1919).

Plantation crops became the focus of attention at Pamplemousses in 1767 when spice plants such as nutmeg (*Myristica fragrans*), pepper (*Piper nigrum*), and cinnamon (*Cinnamomum zeylanicum*) were acquired from the Dutch Indies, as were various fruit, dye, and varnish plants (Hart, 1919; Purseglove, 1959). Before the days of refrigeration, spice plants were in heavy demand to preserve meats; furthermore, heavily spiced dishes mask the flavor of fish or meat on the verge of spoiling. In the late 1770s and early 1780s, Pamplemousses began distributing seedlings of pepper, cinnamon, clove (*Eugenia caryophyllus*), and nutmeg to the inhabitants of Mauritius as well as the French Caribbean and Guiana.

The staff at Pamplemousses were not content just to introduce a crop

[2] The following sources were used to establish the founding dates of botanic gardens, listed here in chronological order and shown in Figure 3.1: 1587, Leyden (MacPhail, 1972:9); 1626, Jardin des Plantes, Paris (Bretschneider, 1935:123); 1670, Edinburgh (Hepper, 1982:32); 1682, Amsterdam (MacPhail, 1972:13); 1735, Pamplemousses, Mauritius (Hart, 1919); 1759, Kew, London (MacPhail, 1972:118); 1766, St. Vincent (Howard, 1954); 1779, Bath Botanical Garden, Jamaica (Eyre, 1966:15); 1787, Calcutta (MacPhail, 1972:105); 1787, Manila (MacPhail, 1972:118); 1796, Penang (Purseglove, 1959); 1811, Jardim Botanico, Rio (Souza, 1945); 1812, Peradeniya, Sri Lanka (Hepper, 1982:128); 1817, Bogor, Indonesia (MacPhail, 1972:103); 1817, Havana (MacPhail, 1972:119); 1819, Trinidad (Hyams and MacQuitty, 1969:67); 1859, Singapore (Burkill, 1918); 1879, Georgetown, Guyana (Howard, 1954); 1898, Entebbe, Uganda (Hyams and MacQuitty, 1969:229); 1926, Lancetilla, Honduras (Permar, 1945).

to Mauritius or other French possessions. Fresh material was constantly being sought to broaden the variability of economic species. Thus in 1769 and 1771 the staff organized expeditions to bring more nutmeg and clove material to the garden. And during the 1860s, excursions undertaken by Pamplemousses resulted in the acquisition of new sugar cane material from New Caledonia and Australia that saved the disease-ravaged Mauritius sugar industry.

The Dutch also established botanic gardens in their overseas possessions to introduce plants and to provision ships. When the Cape colony was founded in 1652, a fruit and vegetable garden was immediately established to help prevent scurvy among crews destined for the Indies or on their way back to Holland (Prest, 1981:48). The Capetown garden, run by the Dutch East India Company, supplied Dutch crews with fresh fruit and vegetables while at port and with sufficient provisions for two weeks at sea. In 1685, Father Guy Tachard (1686:72) observed lemon, pomegranate (*Punica granatum*), orange, apple, pear, apricot, banana, pineapple (*Ananas comosus*), and rare fruits in the garden. The French Jesuit marveled at the care with which the plants had been brought from distant lands and tended at Capetown.

The Dutch established a major acclimatization garden at Bogor (then Buitenzorg) in Java in 1817. It was set up to screen exotic germplasm for plantations in Indonesia and to study the native flora. Bogor, an 87-hectare garden in the city center and an hour's drive from Jakarta, played a leading role in the introduction of African oil palm (*Elaeis guineensis*), which provides a good oil from the olive-sized fruits and kernels for making margarine and soaps, to Southeast Asia. In 1848, Bogor received four specimens of the West African palm: two from La Réunion (then Bourbon), and two from the Amsterdam Botanic Garden (Purseglove, 1975:482). Two of the original four introductions still thrive at Bogor, where they merit special plaques and are enthusiastically pointed out by tourist guides.

Seeds from the four African oil palms were planted in trials and as an ornamental throughout Indonesia. Extensive rows of African oil palm lined roads traversing tobacco estates in Deli, Sumatra, and furnished seeds for the first plantations of the forest palm in that area around 1911 (Hardon, 1976). The Singapore Botanic Gardens obtained African oil palm seeds from Java in 1870 and multiplied the palm for distribution throughout Malaysia, initially as an ornamental. Some of the original African oil palms (Figure 3.2) introduced to the Singapore Botanic Gardens still stand in the northern extension of the gardens.[3] The

---

[3] The northern extension of the gardens is land that originally belonged to the gardens but was taken away for the development of Raffles College in the early part of this century.

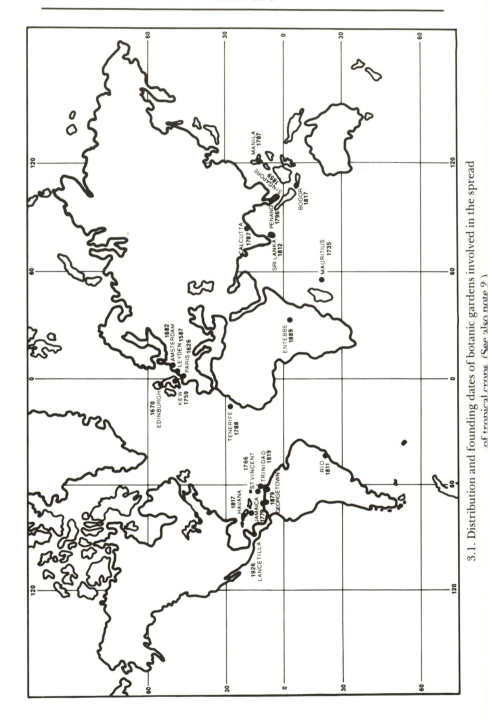

3.1. Distribution and founding dates of botanic gardens involved in the spread of tropical crops. (See also note 2.)

original African oil palm introductions to the Bogor and Singapore gardens came from unimproved stocks and are much taller than modern cultivated varieties, which are relatively short to facilitate harvesting of the fruits.

Bogor and its daughter botanic gardens also worked with other crops. In 1826, Bogor was the site for the first planting of tea (*Camellia sinensis*) in Java, which subsequently became a major crop in highland parts of the island. Jasmine tea, rather than coffee, is the most commonly consumed beverage in Indonesia today. Bogor also obtained rubber (*Hevea brasiliensis* and other latex-bearing trees) and fresh coffee materials to combat rust. By the late 1800s, staff at Bogor were investigating diseases of rice, sugar cane, tobacco, and coffee (*Coffea arabica*). To accommodate the ever-increasing need for space to plant and multiply crop plants, Bogor acquired a 75-hectare satellite garden, Tjikeumeuth, 4 kilometers from the main garden in 1875 (Massart, 1945). This space is now occupied by urban growth and the Institute of Industrial Crops Research.

Under the energetic leadership of Melchior Treub, who assumed the directorship of the Bogor Botanic Garden in 1883, further space for the garden's work was created by forging arrangements with owners of sugar cane, tobacco, rubber, cinchona (*Cinchona* spp.), cacao (*Theobroma cacao*), tea, and coffee estates to relinquish part of their holdings for experiments. Experimental plots remained under the ownership of estates, but research work was directed by the Bogor Botanic Garden (Bernard, 1945). Dr. Treub stressed that the garden should directly serve agriculture and industry as well as conduct basic research; the zoological museum within the garden, for example, studied agricultural pests, among other subjects.

The Portuguese, as in the case of the Spanish, were also active in moving crops between continents, but unlike the Dutch they generally made direct introductions rather than use botanic gardens as staging areas. They were nevertheless responsible for establishing one of South America's foremost botanic gardens, Rio's Jardim Botanico, which has been drawing appreciative crowds since 1811 and is particularly admired for its main avenue lined with majestic royal palms (Souza, 1945).

Rio's Jardim Botanico was established as an acclimatization garden for foreign economic plants. It soon had sizable collections of tea, cam-

---

Raffles College became part of the University of Singapore after independence, and Singapore Botanic Gardens received the land back in 1984 when the university moved to a new campus. The northern extension of the gardens, reached by an underpass from the main section, was the site of many of the economic plant introductions to Singapore. It was there, for example, where rubber trees were multiplied in the early part of this century. The newly reacquired 15-hectare site is being developed to display crop plants.

3.2. African oil palm from Java in the Singapore Botanic Gardens, 1985. These palms are descendants of the four introductions at Bogor.

phor (*Cinnamomum camphor*), cinnamon, nutmeg, breadfruit (*Artocarpus altilis*), jackfruit (*A. heterophyllus*), clove, litchi (*Nephelium litchi*), mango, and carambola (*Averrhoa carambola*). Brazilian and foreign visitors to the garden sometimes complained about the lack of native flora (Gardner, 1846:34; Ribeyrolles, 1941:157; Cruls, 1949:257). Only after 1890 did the Jardim Botanico contain many Brazilian plants (Porto, 1936). Tea was one of the first plants grown at the garden, which was responsible for introducing the crop to South America (Figure 3.3).

The Portuguese had visions that Brazil might one day rival China as a major tea producer. Count Linhares[4] was responsible for bringing tea seedlings from Macao to the Jardim Botanico (then known as the Horto Botanico and renamed the Real Jardim Botanico in 1819); two

[4] Rodrigo de Souza Coutinho, Minister of War and the Exterior.

3.3. Tea being planted by Chinese laborers in the Jardim
Botanico, Rio de Janeiro, in the early nineteenth century.
(From Rugendas, 1941.)

hundred Chinese laborers were also imported to look after the young
tea plants (Cruls, 1949:257). By 1825, six thousand tea plants were
thriving in the Jardim Botanico and seedlings were distributed to São
Paulo, Minas Gerais, and Teresopolis (Ferrez and Mouillot, 1965:106).
In 1836, one plantation in São Paulo had upwards of 20,000 tea bushes,
but the crop never took off in a major way because it proved inferior to
imported Chinese tea and was expensive to produce because of high labor costs (Gardner, 1846:35). Although tea is a profitable crop in certain parts of South America, such as in highland Ecuador, the continent
has never been able to challenge China, India, or Sri Lanka as a major
exporter of the commodity.

The Spanish established several botanic gardens in the tropics and
subtropics, but few have survived to this day. One notable exception is
the Jardim de Aclimatación at the Tenerife port of Orotava in the Ca-

nary Islands. The well-maintained garden was founded by royal order in 1788 and received its first shipment of exotic seeds and plants the same year (J. J. Parsons, pers. comm.). One of the main purposes of the Orotava Jardim de Aclimatación was to adapt tropical plants to European weather conditions (Humboldt, 1818:137).

The British were especially active in establishing botanic gardens in the tropics. The Royal Society of London, for example, promoted the first garden in the New World for the acclimatization of tropical crop germplasm on St. Vincent in the Windward Islands in 1766 (Howard, 1954; Watkins, 1976). The St. Vincent Botanic Garden was the gateway for the introduction of breadfruit into the Neotropics; Captain William Bligh brought six varieties of the starchy dooryard tree, which is native to Southeast Asia and the Pacific, to the garden from Tahiti in 1793. The first attempt to do so in 1789 had failed when the crew of the *Bounty*, which was carrying the breadfruit, mutinied, partly because so much precious fresh water was used to sustain the approximately one thousand young trees on board (Oster and Oster, 1985). Breadfruit suckers were propagated at the St. Vincent Botanic Garden for distribution on the island and to other botanic gardens in the Caribbean. Breadfruit was brought to the New World to provide cheap food for field workers on sugar cane, coffee, cacao, and indigo plantations. At first the new fruit met with some resistance, but it has since blossomed into a regional favorite.

The St. Vincent Botanic Garden also played an important role in the multiplication and dissemination of another Asian tree crop, nutmeg. Seeds of this spice plant were obtained by the botanic garden from French Guiana in 1791 (Howard, 1954). In 1793 the garden probably also held some samples of an important sugar cane variety, Otaheiti, which originated in Tahiti and, along with other introductions of the same variety, helped improve sugar cane production in the Americas during the nineteenth century (Galloway, 1985). For most of its history, the St. Vincent Botanic Garden has been heavily involved in agricultural research. Garden staff, for example, developed one of the world's best cotton cultivars, St. Vincent Superfine Sea Island, as well as highly productive arrowroot (*Maranta arundinacea*) varieties.

The British established several botanic gardens on Jamaica in the eighteenth and nineteenth centuries to test and propagate suitable tropical crops from Asia, the Pacific, and Africa. Governor Sir Basil Keith, for example, authorized the construction of Bath Botanic Garden near Gordon Town in 1779, which introduced the highly successful Cavendish banana to Jamaica. The Carleton Botanic Garden, established on 6 hectares in 1862, released the Bombay mango and navel orange to Jamaica in 1868 and 1870, respectively (Eyre, 1966:22, 29).

Private enterprise was also behind the establishment of several botanic gardens in British overseas possessions. The British East India Company, for example, started a spice garden soon after Penang was settled in 1786. The main purpose of the extensive garden, which officially opened in 1796, was to break the Dutch monopoly on spice cultivation. To that end a botanist, Christopher Smith, was dispatched to the Moluccas in 1796 to collect nutmeg and clove seedlings. Smith sent 71,266 live nutmegs, 55,264 clove seedlings, and several other economic plants to the Penang garden (Ridley, 1910). The same company was also behind the establishment of the Calcutta Botanic Garden in 1787. The 112-hectare garden on low-lying alluvial soil was initially planted to timber trees, such as teak (*Tectona grandis*) and mahogany, for shipbuilding (Holttum, 1984). Research on improving jute (*Corchorus* sp.), Indian cotton, tea, cinchona, coffee, cacao, flax (*Linum usitatissimum*), hemp (*Cannabis sativa*), and cassava soon followed (Sharma, 1984).

The Royal Botanic Gardens at Kew often assisted other botanic gardens within the British Empire by providing staff, advice, and material (Ashton, 1981). Kew was a catalyst in setting up Sri Lanka's botanic gardens in 1812 and helped nurture Singapore's botanic gardens which opened in 1859. Directors and staff of overseas botanic gardens sometimes came from Kew or received their training there.

The Singapore and Sri Lanka gardens had a strong emphasis on economic plants. The 58-hectare Peradeniya garden in Sri Lanka, for example, supplied coffee and rubber seedlings to estates, and a satellite garden, Hakgala, which opened in 1861, propagated cinchona (*Cinchona succirubra, C. calisaya*). In 1876 alone, the Hakgala garden distributed 1.2 million cinchona plants (Rajapakse, 1984). By 1877 the Sri Lanka gardens had distributed 5.5 million cinchona plants to growers supplying the London market (Brockway, 1979:122). Peradeniya stopped working with coffee when rust (*Hemileia vastatrix*) wiped out Sri Lanka's plantations, and Hakgala ceased propagating cinchona when competition from large-scale plantations in India and Indonesia undercut local production. Peradeniya and associated botanic gardens introduced several other crops to Sri Lanka, including the tomato (1814), cacao from Trinidad (1834), durian (*Durio zibethinus*) (c. 1850), camphor (1852), vanilla (*Vanilla planifolia*) (1853), cherimoya (*Annona cherimolia*) (1882), and various forest species (Rajapakse, 1984).

Kew has employed a network approach to making germplasm more available since its inception. Its emphasis on collaboration with other botanic gardens is especially evident in the history of two commercially important plants, cinchona and rubber. In 1860, Richard Spruce and Robert Cross collected seeds of the red bark quinine tree (*Cinchona succiruba*) in forests cloaking the eastern slopes of the Ecuadorian Andes. The bark of cinchona trees contains an alkaloid, quinine, that is toxic to malaria parasites and at that time was the only effective treatment for the debilitating and sometimes fatal disease. Dried seed, cuttings, and seedlings were dispatched to Kew; some of the material somehow survived mule trains, raft trips in hot and humid weather, and a lengthy steamship voyage to England.

Meanwhile farther south, Kew had two other collectors, Clements Markham and Pritchett, gathering seeds of other species of cinchona to be germinated at Kew (Brockway, 1979:114). Robert Cross made further trips to Ecuador in 1861 and to Colombia in 1863 and 1868. Vast numbers of seedlings and cuttings were sent from Kew to India, Sri Lanka, and the West Indies, where quinine tree plantations were soon established. In India, lifesaving and inexpensive doses of quinine subsequently became widely available (Hepper, 1982:131).

The rubber story has become, to some at least, one of the most infamous chapters in Kew's history. Some perceive botanic gardens, such as Kew, as mere instruments in the colonial powers' drive to obtain natural resources from tropical countries. While it is true that botanic gardens were often set up to secure commercially important plants for planting in colonial possessions, this effort has clearly benefited citizens of the Third World. The manner in which rubber was obtained from its native habitat, the Amazon rain forest, has sparked controversy; some have accused Kew collectors of smuggling seeds of the latex-bearing tree out of Brazil (Brockway, 1979:32; Weinstein, 1983:219). And as shown in a previous publication (N. Smith, 1984), one of the authors of this book has also been guilty of this common misconception.

The circumstances regarding the exodus of *Hevea brasiliensis* seeds from Amazonia do not support the popular notion of illicit behavior. In 1876, Henry Wickham and Robert Cross obtained 70,000 rubber seeds in the Amazon; the activities involved in arranging such a vast consignment would not have escaped the attention of Brazilian authorities. Besides, the export of rubber seeds from Brazil was not banned at that time (Wycherley, 1959; Purseglove, 1974:149; Voon, 1976:3). Furthermore, Wickham secured customs clearance for shipping the seeds to England from Santarém (Majid and Hendranata, 1975; Hepper, 1982:131). Rubber seeds do not remain viable for very long and only 2,397 of the seeds sent by Wickham germinated at Kew. Still in 1876, 1,919 of those seedlings were sent to Sri Lanka and planted at Hener-

3.4. One of the two remaining descendants of the original 1877 shipment of rubber (*Hevea brasiliensis*) to the Singapore Botanic Gardens.

atgoda, a low-lying satellite garden of Peradeniya. In the same year, Kew sent 50 rubber seedlings to the Singapore Botanic Gardens but they perished at the docks because of a delay in paying freight charges (Purseglove, 1959). A year later Kew sent a further 22 rubber seedlings to the Singapore Botanic Gardens where they arrived in good condition and thrived. By 1897, 1,310 seed-bearing *Hevea* trees were established at the gardens (Voon, 1976:9). Two descendants of the original shipment were still alive in the garden in 1985 (Figure 3.4).

Spurred by a quickening demand for bicycle tires in the 1880s, and later for automobiles, the Singapore Botanic Gardens served as a nursery for the large-scale development of rubber plantations in the early 1900s and eventually dispatched 7 million seeds and a smaller amount of seedlings (Purseglove, 1974:150). Most of the seedlings were initially distributed in Southeast Asia; as early as 1905, rubber trees were growing on 5.3 million hectares in Britain's Asian protectorates, and by 1915 the planted area had grown to nearly 102 million hectares (Weinstein, 1983:219). By 1902, plantations in Malaysia were virtually self-sufficient in planting stock, so the Singapore Botanic Gardens supplied seeds to other regions (Ridley, 1903:5). In 1910, for example, rubber seeds produced by the gardens were in strong demand in African territories, especially Uganda, Nigeria, and Liberia (Ridley, 1911:5). Rubber seeds were even sent to Honduras and British Guiana, close to the native home of the mottled gray-barked tree, and in 1906, 900 rubber seeds were sent to Mexico and 500 to the Philippines (Ridley, 1907:6).

Kew played a key role in the spread of rubber plantations in the tropics, but botanic gardens in the homelands of other colonial powers have also frequently served as steppingstones for the dispersal of tropical crops. A coffee plant from Java, for example, was taken to the Amsterdam Botanic Garden in 1706; then material was sent to Surinam (then Dutch Guiana) in 1718 (Purseglove, 1974:460). In 1713, the Burgomaster of Amsterdam sent Louis XIV a progeny of the Javan coffee plant that was subsequently tended in the Jardin des Plantes. Progeny of this introduction to Paris were sent to Martinique in 1720, but only one seedling survived the journey to the Caribbean island. It nevertheless started the coffee industry on Martinique and provided material for Jamaica, which led to the development of the premium Blue Mountain variety, considered by some aficionados to be the world's best coffee. The Edinburgh Botanic Garden also obtained offspring from the Java coffee tree in the Amsterdam Botanic Garden and sent material to Malawi (then Nyasaland) in 1878.

Vanilla and cinchona also illustrate the way-station function of botanic gardens in colonial homelands. Leyden Botanic Garden, for example, provided a staging ground for the transfer of vanilla specimens from South America to Java (Hyams and MacQuitty, 1969:43). The first cinchona (*Cinchona calisaya*) brought to Java was initially grown in a hothouse of the Leyden Botanical Garden. In 1851, a single *Cinchona calisaya* seedling was received in poor condition at the Bogor Botanic Garden. Before the seedling succumbed, Dr. Teijsmann took a cutting that subsequently rooted in the strawberry garden of the governor general on the slopes of the Gedeh volcano (Van Gorkom, 1945), which became the Cibodas Botanic Garden, a satellite garden of Bogor, in 1874 (Figure 3.5). Because of its tropical highland location, Cibodas was ideal for propagating cinchona, a native of Andean forests. In 1854, a further five hundred cinchona plants were shipped to Bogor from Leyden, but only seventy-five were still alive when they reached the garden (Van Leersum, 1945). The survivors were sent to Cibodas to be multiplied,

3.5. Cibodas, a satellite garden of Bogor, on the slopes of
Gedeh volcano, Java, 1985.

and that garden consequently provided initial planting stock for cinchona plantations on Java, though that function was soon taken over by
the plantations themselves. Java eventually provided 80 percent of the
world market for cinchona bark (Brockway, 1979:120), and in 1985
several mature cinchona specimens were still growing at the cloudy and
moist garden.

Botanic gardens in the United States are generally younger than
those in the Old World, the Caribbean, and Latin America. A few pri-
vately owned botanic gardens, chiefly concerned with medicinal plants,
were established in Pennsylvania in the early 1700s (Earnest, 1940:17);
but major botanic gardens with research staff were not established in
the United States until the latter half of the nineteenth century. For example, the Missouri Botanic Garden in St. Louis, the Arnold Arboretum in Boston, and the New York Botanic Garden were founded in
1859, 1872, and 1894, respectively (Fairchild, 1938:136). Traditionally

these gardens have concentrated on wild species and ornamentals rather than crop plants. It was not until 1898 that the U.S. Department of Agriculture enacted a plant-introduction system for economic and ornamental plants (Cunningham, 1984:268). Known as the Office of Foreign Seed and Plant Introduction and headquartered in Washington, D.C., the service used greenhouses on the grounds of the Capitol as well as plant introduction gardens or stations throughout the United States. By 1910, five plant-introduction stations were operating, including one in Chico, California, another in Miami that specialized in tropical plants, and a third in Brooksville, Florida, which was used for testing and maintaining bamboo and taro (*Colocasia esculenta*).

Two main developments led to the establishment of the Office of Foreign Seed and Plant Introduction. The political defeat of cattle ranchers in the northwest plains in the late 1800s opened up the region to crop farming and created a need for cold-tolerant crops (Fairchild, 1938:114). And in the Southwest, farmers were besieging the U.S. Department of Agriculture with requests for drought-tolerant plants. Although the Office of Foreign Seed and Plant Introduction was set up to service the needs of American farmers, it cooperated with other governments and institutions; for example, it promoted the free exchange of germplasm between U.S. plant introduction stations and foreign botanic gardens and nurseries.

Private companies have set up botanic gardens and have played key roles in the collection and introduction of useful plants. Commercial nurseries in England have been selling tree seedlings, turf, and flowering plants since the fourteenth century (Harvey, 1981:17). For example, in 1787, the British East India Company founded the Calcutta Botanic Garden for the acclimatization of tropical crops (Hyams and MacQuitty, 1969:220). And the West Indian Gardens, owned by Mr. F. O. Popenoe in Altadena, California, financed collecting trips to Central America around the turn of this century; the nursery was instrumental in introducing the avocado (*Persea americana*), a neotropical fruit, to the table of Californians with the release of the cold-tolerant Fuerte variety in 1911 (Schroeder, 1967).

Another American company, United Fruit, established a plant-introduction garden and experiment station at Lancetilla near Tela on the northern coast of Honduras in 1926. Wilson Popenoe, the son of the owner of the West Indian Gardens, was picked to run the station, which was charged with screening candidates to replace banana plantations ravaged by Panama wilt disease. The Lancetilla station tested banana clones for resistance to Panama disease and sought other crops that might prove profitable in the American tropics.

Lancetilla, administered by various agencies of the Honduran government since 1974, has made several outstanding contributions to agriculture in the American tropics. Banana plants resistant to Panama wilt disease were found in the Saigon Botanic Garden in 1925, and suckers were first taken to a quarantine station on an island off northwest Panama and then to Lancetilla for further testing in 1928 (Dunlap, 1967). These clones were used to develop Valery, a variety that helped reestablish the banana industry in Central America. In 1926, Lancetilla obtained seeds of African oil palm from the United States Rubber Company in Sumatra and from other sources in Java and Malaysia (Permar, 1945:15). The station subsequently released commercial cultivars of the palm in Central America during the early 1940s. Lancetilla introduced several high-yielding citrus varieties to Central America and has the largest collection of Asiatic fruit trees in the New World.

Although botanic gardens have long played a crucial role in introducing crops to new regions and conserving some plant germplasm, these important functions have recently dwindled. Lancetilla, for example, like many botanic gardens and plant-introduction stations, has seen its heyday. In 1965, thirteen hundred species were maintained on the 435-hectare station, but some three hundred species were lost because of inadequate care between 1965 and 1978. Many tropical botanic gardens are poorly maintained and are little more than public parks (Ashton, 1981; J. Sauer, pers. comm.). Other botanic gardens have succumbed to urban development or have folded due to lack of government support. Governments in developing countries often perceive botanic gardens to be no longer vital to agriculture and plant introduction and have therefore severely cut their operating budgets.

Activities involved in acquiring crop germplasm and in selecting varieties for release to farmers have now largely been taken over by agricultural organizations operated by governments, international agencies, and private companies. The Peradeniya Garden in Sri Lanka, for example, stopped research on food and industrial crops in 1912 when the Ministry of Agriculture assumed that task (Rajapakse, 1984). Pamplemousses was absorbed by the Mauritius Agricultural Department in 1913 (Purseglove, 1959). The Singapore Botanic Gardens ceased work on economic plants in 1925 and lost nearly half its land to Raffles College at the same time (Purseglove, 1959). Garden staff currently do most of their research on tissue culture of orchids.

Most of the remaining botanic gardens in the Third World serve as popular recreation sites for the public. The Bogor and Cibodas gardens, for example, are greatly appreciated by the public; approximately half a million Indonesians and tourists visited the Bogor garden in 1979

3.6. A well-maintained palm group at the Singapore
Botanic Gardens, 1985.

(Sastrapradja and Prana, 1980), and the number of visitors climbs each
year. The well-manicured Singapore Botanic Gardens (Figure 3.6) are
also popular with the general public, especially picnickers, joggers, and
early morning tai chi practitioners.

PLANT HUNTERS

No discussion of crop movement and germplasm conservation would
be complete without examining the work of individuals who risked
their health and lives in search of new crop candidates. Plant hunters
have played crucial roles in the history of plant germplasm collection
and preservation. The ranks of plant collectors through the centuries
include professionally trained botanists, geneticists, physicians, gardeners, explorers, missionaries, and consular officials. Often the importance of these collectors is overlooked or forgotten, yet some species
and many varieties would now be extinct without their efforts (Kingdon
Ward, 1924:19). Few collectors left any printed accounts of their trav-

els, and many were amateur naturalists with little chance of publishing in scientific journals. Furthermore, tracking the outcome of individual plant introductions is sporadic, making the accomplishments of plant explorers often obscure.

Plant-collecting expeditions have been mounted for over four thousand years. Around 2500 B.C., for example, the Sumerians dispatched plant collectors to the heart of Asia Minor in search of vines, figs, and roses (Woolley, 1930:79; Klose, 1950:3). A thousand years later, on the return of an expedition that had gathered incense trees in East Africa, Queen Hatshepsut of Egypt had the event depicted on the walls of a Temple at Thebes (Ryerson, 1933). Other ancient civilizations surely organized similar expeditions, but they left little detailed information about the size of the collecting parties, the sites visited, and the plants brought back.

It was not until the age of discovery in the sixteenth century that plant hunting started a trend that would revolutionize the world. Beginning in the late 1500s, plant-collecting expeditions were sent to more places and covered more territory than ever before. Between 1570 and 1577, for example, Francisco Hernandez, a Spanish physician, sent seeds and live plants from Mexico to the Royal Botanic Gardens at Aranjuez near Madrid (Steele, 1964:6). And John Tradescant, who died in 1638, was one of the first persons to organize plant-collecting trips on a systematic basis. He became a gardener to Sir Robert Cecil, later the Duke of Buckingham, until he was appointed keeper of the garden of King Charles I. Tradescant collected plants in France, Holland, Russia, and Algeria. He took the larch tree from Russia to England, and lilac, crocus, and jasmine from the Mediterranean to Britain. His son soon followed in his footsteps, traveling to Virginia and the West Indies.

Another early plant collector who made extensive trips overseas was Sir Joseph Banks. On behalf of the Navy and the Royal Society, Joseph Banks joined Captain James Cook on a voyage to Tahiti to track the transit of Venus. In addition to Tahiti, Banks and Cook visited Australia (where Banks named Botany Bay), New Zealand, New Guinea, and the East Indies, returning home via the Cape of Good Hope. Banks became the unofficial director of the Royal Botanic Gardens at Kew, which became a major facility for moving plants around the world. British botanists, many of whom became famous for their courage and accomplishments, were sent all over the world. Richard Spruce, George Forest, and Frank Kingdon Ward emerged as giants of the plant-collecting world in the nineteenth and twentieth centuries.

The United States did not undertake government-sponsored introductions until after the War of 1812, though Benjamin Franklin sent seeds and plants to the United States during his European visits in the latter half of the eighteenth century (Hyland, 1977, 1984). But it was not until 1854 that the first plant explorer employed by the U.S. government, D. J. Brown, was sent abroad, in this case to Europe to obtain seeds (USDA, 1971:10). In 1858, the Commissioner of Patents hired Robert Fortune to go to China and collect tea seeds with the intention of eventually establishing tea plantations in the southern United States (Hyland, 1977). Following the establishment of the U.S. Department of Agriculture in 1862, plant exploration was given a further boost. In 1864, for example, an American plant explorer was sent to China to study and collect sorghums, and, until 1898, U.S. government-sponsored collecting missions throughout the world brought back numerous accessions of navel orange, flax, olive (*Olea europea*), persimmon (*Diospyros kaki*), cereals such as wheat and sorghum, and fruit trees.

One of the world's outstanding plant explorers was brought into the service of the U.S. government in the early part of the twentieth century. Frank Meyer, a restless character with an insatiable thirst for knowledge about plants, started work as a gardener in Amsterdam, where he was born in 1875. Thanks to the painstaking work of Isabel Shipley Cunningham (1984), Meyer's contribution to agriculture and gardening has been carefully documented. An incessant wanderlust drove Meyer to emigrate to the United States in 1901, where he was eventually hired as an agricultural explorer by David Fairchild, first director of the Office of Foreign Seed and Plant Introduction. He was sent to China three times, as well as to Europe, Russia, and Tibet. Unlike some of his British contemporaries, such as E. H. Wilson, he sought hardy crops and their wild relatives as well as ornamentals. Meyer was particularly interested in crops that tolerated cold, drought, and alkaline soil.

During his thirteen-year career with the Office of Foreign Seed and Plant Introduction, Meyer was responsible for 2,500 plant introductions to the United States. The intrepid explorer's contribution would surely have been greater had he not died while working in China in 1918. Progeny of his introductions are growing not only in botanic gardens and arboretums today, but are found in private gardens and farmers' fields as well. In Manchuria, for example, Meyer collected seeds of a large-leafed spinach (*Spinacia oleracea*) that were used to develop a variety called Virginia Savoy that resisted blight and wilt, diseases that threatened the spinach-canning industry in the United States. And in China, the naturalized American plant hunter obtained stones of a northern wild peach (*Prunus davidiana*) that proved resistant to root-knot nematodes and has been extensively used as a root stock for grafting apricots, plums, and peaches (Cunningham, 1984:50, 263).

Meyer's introductions have also proved useful for windbreaks and as ground cover. Between 1935 and 1942, for example, some 17,000 miles of Siberian and Chinese elms (*Ulmus pumila* and *U. parvifolia*), species gathered by Meyer on his trips, were planted in the Great Plains to control wind erosion. In 1911, Meyer picked seeds of *Coronilla varia* near Saratov in the Soviet Union that were subsequently used by the U.S. Soil Conservation Service and the Iowa Agricultural Experiment Station to develop emerald crownvetch. This drought and cold-tolerant legume has been widely planted along the banks of interstate highways in the northern United States to check soil erosion (Cunningham, 1984:261, 142). And one of Florida's best lawn grasses, centipede (*Eremochloa ophiuroides*), was collected by Meyer during his last trip to China (Fairchild, 1938:456).

Two other agricultural explorers employed by the U.S. Department of Agriculture, David Fairchild and Wilson Popenoe (Figure 3.7), were also busy acquiring economic plants around the turn of the century and in the early 1900s. David Fairchild, Meyer's boss, collected around the world on foot and horseback. Fairchild, after whom the Fairchild Gardens in Miami are named, sent desirable mango cultivars from Asia to Florida and an alfalfa from Peru that was adopted by farmers in southern California (Fairchild, 1938:128). Between 1916 and 1917, Popenoe sent twenty-four Guatemalan avocado varieties to the United States, many of which proved useful in California and Florida[5] where numerous plantations of the crop were established (Fairchild, 1938:453). The extensive and continuous free distribution of mango and avocado germplasm by the U.S. Department of Agriculture is a lasting legacy to the efforts of Fairchild and Popenoe. A letter of credential issued to Wilson Popenoe in 1916 by Carl Vrooman, Acting Secretary of Agriculture, exemplifies the spirit in which such collecting expeditions were made:

Be it known that Mr. Wilson Popenoe, Agricultural Explorer, in the Office of Foreign Seed and Plant Introduction, Bureau of Plant Industry, of the United States Department of Agriculture, a citizen of the State of California and of the United States of America, has been authorized by me to make a study of the wild and cultivated plants of the Republics of Guatemala, Honduras, Nicaragua, El Salvador, and Costa Rica, for the purposes of ascertaining in what way a mu-

[5] *Plant Immigrants*, No. 197, p. 1802, September 1922 (Office of Foreign Seed and Plant Introduction, Washington, D.C.).

3.7. Wilson Popenoe in Guatemala, c. 1920. (Courtesy of
the Hunt Institute for Botanical Documentation,
Carnegie-Mellon University, Pittsburgh.)

tually beneficial exchange of seeds and plants may be brought about between
those Republics and this country. (Sola, 1967)

Another outstanding and courageous plant collector, Nikolai I. Vavilov (Figure 3.8), had perhaps the broadest experience of any collector
until that time. Vavilov conducted lengthy overland plant-hunting trips
in the U.S.S.R., and in over 50 countries in Asia, the Americas, northern Africa, Europe, and the Mediterranean during the 1920s and
1930s (Popovsky, 1984:2). He was a geneticist by training and is well
known for his ideas on centers of plant diversity and crop domestication. He organized expeditions that, together with exchanges from
other institutions, amassed over 50,000 seed samples of wheat, rye,
oats, peas, lentils, beans, chickpeas, and maize (Vavilov, 1957:2). This

3.8. Nikolai Vavilov (right) with Roger P. V. de
Vilmorin, c. 1925. (Courtesy of the Hunt Institute for
Botanical Documentation, Carnegie-Mellon
University, Pittsburgh.)

large collection of crop plants and close relatives from afar provided the
foundation for the establishment of modern gene banks in the Soviet
Union.

Vavilov's concern with wild relatives of economic plants was far-
sighted, since wild species are usually poorly represented in modern
gene banks in spite of their demonstrated usefulness in plant breeding.
He also emphasized sampling the entire range of a species in order to
gather as much genetic diversity as possible. Vavilov drew attention to
the economic potential of germplasm collections, particularly with regard to adaptation and disease resistance (Hawkes, 1978). As a field naturalist, he also recognized the importance of taxonomic studies. During
his travels to Central and South America in 1925 and 1932, the Russian

63

scientist obtained numerous samples of maize and potato and their wild relatives. He personally collected over 60,000 plant samples, including 2,800 specimens of maize (Wilkes, 1972; Popovsky, 1984:161).

Whereas suicide or foul play ended the restless life of Frank Meyer in 1918 (his body was found in the Yellow River in China), it was differences with state officials that brought Vavilov's brilliant career to a halt. He was the first director of the Institute of Applied Botany and New Crops in Leningrad, and his frequent travels abroad aroused suspicion. The Soviet government suppressed scientific plant-collecting expeditions to foreign lands, and, as a consequence, Vavilov never left the Soviet Union after the spring of 1933. Differences with government officials finally reached the stage where his work was severely constrained. He was arrested in 1940 and charged with espionage and efforts to harm agriculture. He died in Saratov prison in January 1943 under extreme conditions (Popovsky, 1984: 191; Kahn, 1985: 108).

Despite the fact that Vavilov had fallen out of favor, many Russian scientists treasured his germplasm collections. During the Second World War, for example, the scientist in charge of the collections at Leningrad reputedly died of starvation rather than touch Vavilov's samples (Evans 1975). Nonetheless, Vavilov was later vindicated, and the Soviet government honored him posthumously by renaming the gene bank the N. I. Vavilov All-Union Institute of Plant Industry. Known worldwide by the abbreviation VIR, the institute remains, because of Vavilov's work, one of the major nodes in the international network of crop germplasm research.

Missionaries have long helped collectors and have actively acquired plant germplasm themselves. Plant explorers in foreign countries typically consulted with the clergy for information as to the most fruitful collecting sites. Missionaries also sometimes collected or forwarded promising material to their home countries. At the beginning of the eighteenth century, for example, Jesuits in China started sending live botanical specimens and seeds to Europe. Material sent by the Catholic fathers was generally taken by traders traveling in caravans through Russia (Bretschneider, 1935:45). Father Jean Pierre Armand David, a member of the Order of St. Vincent de Paul, went to Peking in 1863 to teach science at a boy's school. An avid naturalist, Father David was soon making trips to the countryside to collect plant specimens and to observe nature. In addition to seeds for planting, Father David sent botanical specimens of approximately 2,000 species to France. In a journal entry made in 1866, he reconciled his priestly duties and his passion for natural history thus: "All species concerned with the works of creation increase the glory of their Creator, for to know truth is to know God" (Healey, 1975:105).

In Chile, Catholic priests obtained alfalfa seeds, which were sent to California where they were extensively planted in the late 1800s (Fairchild, 1938:128). A missionary was responsible for the introduction of the King of Siam orange into Florida (Fairchild, 1938:89). Reverend F. Schneider, a Presbyterian missionary in Bahia, Brazil, sent twelve newly budded orange trees of a seedless variety to the U.S. Commissioner of Agriculture in 1869. Samples were soon sent to Riverside, California, where they were propagated on a large scale and became the highly successful cultivar, Washington Navel. By 1921, Washington Navel accounted for the bulk of orange-tree groves in California (Klose, 1950:78).

The diplomatic services of countries seeking foreign plant germplasm have also aided the work of plant collectors and have occasionally secured promising material. While Thomas Jefferson was Minister to France between 1784 and 1789, for example, he sent seeds of grasses, cereal crops, vegetables, and cuttings of olive and fruit trees to correspondents and nongovernment organizations in the United States (Ryerson, 1967). In 1819, the U.S. Secretary of Treasury issued a circular pointing out the importance of crop plants and requesting that consuls and naval officers send home useful plants. Eight years later, President John Quincy Adams had another circular printed reiterating the 1819 statement, but this time giving explicit labeling and packaging instructions for plant material. Not until 1839, however, did the United States government provide direct financial support for the acquisition of foreign plant germplasm. The modest program for plant introduction was housed in the Patent Office of the Department of State. Prior to the Civil War, the plant-introduction program in the United States focused on tea, cereals, vegetables, cinchona, date palm (*Phoenix dactylifera*), olive, cork oak, camphor, cotton (*Gossypium* spp.), and alfalfa (Ryerson, 1967). Paradoxically, the United States has the world's most comprehensive collection of date palms, and germplasm of the palm is routinely supplied to countries that should have established their own collections.

The diplomatic services of other countries have also been directly involved in plant introductions. Dr. Augustine Henry, an Irish consular official in China, for example, sent plant samples to his homeland in the latter part of the nineteenth century, and Frank Meyer sent several specimens to the United States via the diplomatic pouch (Cunningham, 1984:44).

The conditions under which plant hunters operate are often trying.

Hardships were particularly severe before the age of air travel and refrigeration. Plant collectors frequently had to endure severe weather in the open; Meyer, for example, would walk all day in the rain with the thermometer approaching the freezing mark. He was often on the road before dawn and would think nothing of sleeping outdoors in subzero temperatures. In an age before vaccines were available, agricultural explorers had to work in areas where hygiene was poor and where epidemics of potentially fatal diseases such as cholera were common; occasionally they were even menaced by thieves or threatened by soldiers. Communications were slow and much material perished on long voyages back to the homeland. Collectors of the caliber of Meyer and Kingdon Ward were able to confront numerous hardships during their expeditions because they were motivated by an insatiable curiosity about plants and foreign lands and because they yearned to see the fruits of their labors planted along roadsides, and in orchards, home gardens, and farmers' fields.

Although modern plant collectors still encounter hardships, they generally have an easier time. Communications are usually quicker, and collectors often work in teams rather than as individuals. Groups are typically composed of several nationalities, and the country in which collections are being made is almost always represented so that expeditions are mutually beneficial. When host countries are involved in plant-collecting trips, they learn more about their germplasm resources and can help arrange export permits. Furthermore, today more material survives the hazards of transportation, thanks to automobiles, refrigeration, plastic bags, and jet aircraft.

MODERN GENE BANKS

The need to conserve forest and animal resources was taught and decreed in parts of China and India as far back as 700 B.C., but genetic conservation as practiced today is a relatively young venture. Modern genetic conservation partly derives its roots from the pioneering studies of Alphonse de Candolle (1855, 1902) and Vavilov (1940, 1957). Following these studies, germplasm collections for exploitation were established in the Soviet Union, the United States, and a few other countries. These collections were primarily for the use of plant breeders and scientists from associated disciplines. Many collections, such as Vavilov's in Leningrad, have been used as research material to elucidate evolutionary processes and taxonomic relationships, identify centers of origin and diversity, discover distribution patterns of significant variability or of particular characteristics, and relate plant variations to the environments that helped shape them.

Evolution in the management and maintenance of crop germplasm has been gradual for most of farming history. But recently the nature of gene banking has changed dramatically. If the time span of agriculture is telescoped into an hour, germplasm has been conserved in botanic gardens and glasshouses only within the last minute, and gene banks with deep freeze capability did not come on the scene until the last few seconds. Furthermore, crop germplasm conservation has grown from individual country efforts to an international amalgam of scientists, governments, commercial enterprises, and private foundations.

Faced with actual or imminent loss of material in the field, scientists spurred the development of modern gene banks because they needed a constant and reliable supply of germplasm readily at hand. It was too costly to organize collecting trips each time they wanted fresh genes to enrich their breeding pools. And breeders came to realize that working collections were no longer sufficient because they could not count on replacing germplasm in the event that material was lost (Mengesha, 1984). All too often plant-collecting expeditions were frustrated when they returned to sites to re-collect materials only to find in their place a village, highway, or reservoir.

The technology for long-term preservation of seeds was in place long before the emergence of modern gene banks. Machines for making ice and for freezing meat were in use by the mid-1800s; this technology, for example, allowed the beef industry to expand in Argentina and to send frozen carcasses to Europe. In the 1920s, freon-based refrigeration equipment was developed, providing a more efficient and less dangerous way of storing goods; earlier systems relied primarily on ammonia, which could leak toxic fumes. Plant breeders and the seed trade were behind the push to store plant germplasm under long-term conditions; fortunately, technology was available when the need arose.

Currently three main types of germplasm collections can be found at agricultural research institutes. Working collections, which are grown out by breeders every year, are kept at ambient temperature or in air-conditioned rooms. In medium-term storage, seeds are dried and kept at $0°$ to $-5°$ C; under such conditions most accessions can remain viable for approximately ten to thirty years. In long-term storage, samples are dried and sealed in airtight containers at $-20°$ C. Collections in long-term storage are rarely disturbed and should last for several decades, in some cases even for over one hundred years. Tests are periodically made to ensure that samples are still capable of germinating. Some agricultural institutes maintain germplasm with all three storage methods, but short- and medium-term collections are more common.

Germplasm of root crops is stored either in field plots or in tissue culture. In the case of potatoes, clones must be planted out every year if they are not held *in vitro* (N. Smith, 1983a). This procedure is time consuming and expensive, and the chance of mislabeling samples increases each time the material is regenerated. With current technology, root crops can be maintained in test tubes for up to two years before cultures must be renewed. Long-term storage of *in vitro* plantlets in liquid nitrogen (cryopreservation) is in the experimental stage and is likely to prove feasible for some species within a decade.

The Soviet Union gained an early lead in storing plant genetic resources, albeit not under ideal conditions at the time, due to the foresight of Nikolai Vavilov, who set up the All-Union Institute of Plant Industry in Leningrad in the 1920s. Accessions at the institute were kept at ambient temperature in metal cases and had to be grown out every year (M. Popovsky, pers. comm.). Many of Vavilov's collections are still maintained at this institute, which acquired the capacity for long-term seed storage in the 1970s.

Crop germplasm has been collected and evaluated in the United States since the last century, but it was not until the 1940s that centers for storing plant germplasm under medium-term conditions were established. Four regional plant-introduction stations were created in the 1940s to maintain germplasm as living collections and as seed in cool storage; these centers, which had mandates to introduce, multiply, evaluate, distribute, and preserve plant germplasm, were located in Ames, Iowa (1947), Geneva, New York (1948), Experiment, Georgia (1949), and Pullman, Washington (1949). An important factor in the decision to include cold storage facilities at the regional introduction stations was the fact that only 5 to 10 percent of the 160,000 plant accessions recorded since 1898 could be accounted for in living collections (Hyland, 1977).

Another major center for storage of crop germplasm, the Inter-Regional Plant Introduction Station for Potatoes, was established at Sturgeon Bay, Wisconsin, in 1949. The first national facility for preserving seeds, the National Seed Storage Laboratory (NSSL) operated by the U.S. Department of Agriculture, was built at Fort Collins, Colorado, in 1958 (Simmonds, 1979:334). The NSSL stored collections of all major crops from around the world, including wheat, oats, barley, maize, sorghum, rice, soybean, flax, tobacco, and cotton at 2° C until, as in the case of the Vavilov Institute, facilities were upgraded for long-term storage in the early 1970s. The American Seed Trade Association was strongly behind the push to upgrade NSSL's facilities for long-term storage (Q. Jones, pers. comm.).

Apart from the international germplasm collection at Gatersleben in the German Democratic Republic, gene banks in Europe, Canada, Japan, and Australia, built up largely through plant introductions, are generally smaller and less comprehensive than those in the United States and the Soviet Union (Williams, 1984a). Countries understandably give higher priority to germplasm useful to their own farmers. Thus Australia concentrates on collections of wild plants for its forage-improvement research programs.

Although gene banks in industrial countries have always assisted plant breeders in the Third World, an awareness began to emerge in the 1960s that more collecting and preserving of the genetic resources of tropical and subtropical crops were needed. The United Nations Food and Agriculture Organization (FAO) spearheaded initial efforts to bring the issue of germplasm conservation to the attention of the world community. In 1959, the FAO published *Plant Introduction Newsletter* No. 6, containing a world list of germplasm collections then in existence, as well as their custodians, which facilitated seed exchange and plant introduction (Wilkes, 1983). In 1961, the FAO organized the first international technical meeting on plant exploration and introduction, and a panel of experts on the topic was established four years later. Two further international technical conferences on crop genetic resources in 1968 and 1973 made proposals for the exploration, collection, conservation, documentation, and evaluation of crop genetic resources and recommended that a global network of crop gene banks be established (Frankel and Bennett, 1970; Frankel and Hawkes, 1975). Echoing this concern, the 1972 United Nations Conference on the Human Environment adopted a resolution calling for an international program for preserving the germplasm of tropical and subtropical crops. Sir Otto Frankel, chairman of the FAO panel, was a catalyst in the early international movement to advance germplasm work.

In the same year, the Consultative Group on International Agricultural Research (CGIAR), which is funded by forty donors including governments, multilateral lending agencies, and private foundations and which supports a network of international agricultural research centers (Figure 2.4), convened a working group at Beltsville, Maryland. It strongly urged the creation of a network of nine regional genetic resource centers and a series of crop-specific institutions in developing countries (Frankel, 1975). The International Board for Plant Genetic Resources (IBPGR), established in 1974 within FAO's headquarters in

Rome, was one of the autonomous international agricultural research centers launched by CGIAR.

When IBPGR was created, it was given a daunting task: the development of a world network of plant genetic resource activities into which all ongoing programs would be articulated (Williams, 1984b). But few programs were properly organized, and priorities were defined only in the broadest terms. IBPGR quickly assumed a central part in stimulating the development of crop germplasm storage facilities in the Third World, despite its relatively small staff and modest budget. In 1985, IBPGR had only twenty-five professional staff members and an annual budget of $4.6 million.

IBPGR provides start-up money for germplasm work, mainly for crops of worldwide or regional importance, but it does not operate its own facilities (Wilkes, 1983). The Board has established joint advisory committees for the crops that receive attention by sister centers within the CGIAR and culls the knowledge of dozens of the world's top scientists. IBPGR defines priorities for collecting trips and monitors progress in conserving the genetic resources of food crops (IBPGR, 1980, 1984a). One of IBPGR's most important tasks is to fund plant exploration, and during its first decade of existence the Board was involved in three hundred collecting missions in eighty-eight countries. These collections spanned 138 species and were placed in gene banks of 450 organizations in ninety-one countries (TAC, 1985:31). The Board has provided equipment for gene banks in over twenty countries and supports work to develop germplasm collections of vegetatively propagated crops. About half of the gene banks supported were for long-term conservation, and these have been incorporated into a network of base collections.[6]

Information about germplasm collections is essential if breeders are to learn of potentially valuable resources for their breeding programs (S. Smith, 1984). IBPGR thus supports efforts to better document gene-bank accessions. The Board accomplishes this goal through several avenues: by defining standardized descriptors on a crop-by-crop basis, printing gene-bank catalogs, and helping gene banks computerize their holdings. IBPGR has also published thirteen directories of germplasm collections that contain information on the size and location of collections, storage conditions, and degree of evaluation of all major crops. It has contributed to both hardware and software of gene banks, having

[6] Information on the extent and crops involved in IBPGR's network of base collections is provided in Chaper 6.

supplied twenty-two microcomputers and associated equipment to twenty centers.

Another of IBPGR's important tasks is supporting and planning for the training of germplasm specialists as well as conferences and workshops on plant genetic resources (Hawkes, 1983:137). To buttress germplasm collecting and conservation efforts in developing countries, the Board has supported training to the Master of Science level of many scientists for the Third World and supports a specifically designed international training course in genetic conservation at the University of Birmingham in the United Kingdom. The University trained 196 individuals, mostly from developing countries, in the collection and management of plant genetic resources between 1969 and 1985. In addition, over six hundred trainees have attended short technical courses in many parts of the world. The Board has sponsored training courses on germplasm documentation alone for more than a hundred people. An intern scheme at the pre- and postdoctoral levels has recently been initiated. IBPGR-sponsored activities cover more than fifty crops and several hundred wild species and involve scientists in about 110 countries.

During IBPGR's first few years of existence, it did not support research except in the development of conservation methods. Now, having established a viable network of gene banks and cooperating scientists, IBPGR intends to place greater emphasis in areas of research related to germplasm storage and evaluation.

4

GENE BANKS

How does a gene bank operate? What are the procedures followed in germplasm collection and storage? And what are the hazards involved? In this chapter we address these questions by outlining the principles guiding the acquisition of plant samples for preservation, describing the functioning of a typical gene bank, and analyzing the problems encountered in storing germplasm. We also explore solutions to difficulties encountered in germplasm storage and describe types of germplasm collections, including those storing orthodox seeds. We go on to discuss the complementary roles of field gene banks and botanic gardens in maintaining economic plants that produce recalcitrant seeds, which cannot be stored by conventional means. Finally, we tackle the issue of whether crop germplasm should be stored *in situ* in farmers' fields and wild species in reserves, instead of *ex situ* in gene banks or field plantings.

GENE BANK PRINCIPLES AND OPERATIONS

Three principles generally guide the collection, conservation, and exchange of germplasm. First, when an accession is gathered, a sample is left in the country of origin for national use. If no suitable storage facilities exist in that country, duplicate material is usually stored elsewhere until it can be safely returned to the country of origin. Second, germplasm is made freely available to all workers who can effectively use it, including germplasm specialists, breeders, and other scientists. Finally, all long-term collections are duplicated and maintained in other locations for safety reasons.

To be useful to plant breeders, a gene bank must have easily retrievable and understandable information about the seeds or plants in stock. The first step in germplasm conservation work, then, is to list characteristics of the plant in the field, describe the environment in which it grows, and note its location. Traits such as seed color or fruit shape are noted, as well as the latitude and longitude of the collecting site. If the list of plant and site characteristics becomes too long, however, the pace of collecting and data recording slows. Field notes on a germplasm sample, which often consist only of the local name, date, and site data where collected, are referred to as "passport data". Before treatment for storage, the accession is numbered (Chang, 1976b:14).

The number of seeds sufficient to preserve the variability of germ-plasm is still disputed, but the larger the sample the greater the chance that rare and potentially useful genes will be included. Frankel and Soulé (1981:34) suggest that 100 seeds are sufficient to cover 99.5 percent of the genetic variance of a population, whereas Yngaard (1983) asserts that a minimum of 250 seeds is required to represent a population. A much larger number is necessary according to Hawkes (1982), who asserts that at least 2,500 seeds are required to capture the genetic diversity of a population.

To be on the safe side, most accessions contain several thousand seeds. At the National Seed Storage Laboratory (NSSL) in Fort Collins, Colorado, when possible, at least 10,000 seeds are maintained per accession for small-seeded species and 5,000 per accession for larger-seeded species (Bass, 1984). At the Gatersleben gene bank in the Democratic Republic of Germany, technicians strive to maintain about 10,000 seeds per accession, which in the case of dried cereal grains weighs approximately half a kilo (Lehman, 1979). At the International Maize and Wheat Improvement Center (CIMMYT—Centro Internacional de Mejoramiento de Maiz y Trigo), maize accessions in long-term storage contain between 5,000 and 17,000 seeds, depending on the variety. In CIMMYT's medium-term wheat gene bank, one-kilogram packets of seed contain about 20,000 seeds. Accessions in long-term storage at the International Rice Research Institute (IRRI) in the Philippines contain between 5,000 and 8,000 seeds. Rice samples in medium-term storage at IRRI contain between 20,000 and 32,000 seeds, depending on the cultivar and species. Accessions in medium-term storage at IRRI and medium-term collections of maize at CIMMYT are larger in order to facilitate exchange, evaluation, and use by breeders. Accessions should be still greater for wild species and traditional varieties because they are more heterogeneous than modern cultivars. For long-term storage, the International Board for Plant Genetic Resources (IBPGR) recommends a minimum of 3,000 seeds for genetically uniform plant populations and at least 4,000 seeds for heterogeneous material (Hanson et al., 1984:3).

Plant material destined for germplasm storage must be multiplied if only a small sample was collected. This is a costly process, and some particularly small samples require as many as five regeneration cycles to produce enough seed for storage. Accessions are also increased so that duplicates can be sent to other gene banks and samples can be provided to breeders (Figure 4.1). Material is also multiplied before storage if it arrives in poor condition, since seeds destined for storage should be of high quality and at maximum viability. To control insects for short-term

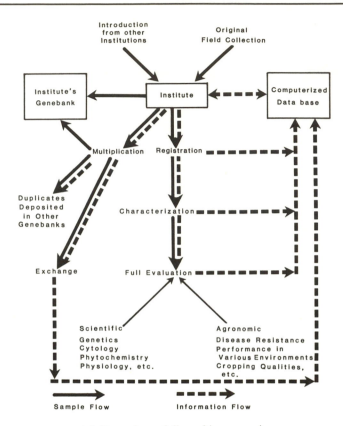

4.1. Procedures followed in processing
a gene-bank accession.

storage, accessions are fumigated or mothballs (diazinon or carbo-furan) are added (Chang, 1983b). For long-term storage, on the other hand, seeds are best left untreated since insects are unlikely to survive temperatures as low as $-20°$ C.

Once germplasm destined for storage has been catalogued, and mul-tiplied if necessary, it should be promptly characterized and evaluated. Checking an accession for desirable characteristics is best done by a team of scientists. This step is particularly important from the stand-point of breeders (Frankel, 1977; W. L. Brown, 1982). A gene bank should not be regarded as a plant museum where relics of the past are merely preserved or displayed. Accessions should be *used*, and breeders need to know what the packets or bottles of seeds on the shelves contain.

An accession should be screened for yield performance and other traits in the presence of pests and diseases. In addition to glasshouse and field testing of accessions, gene-bank material often needs to be scrutinized in a laboratory to uncover its nutritional value, cooking properties, and other characteristics.

Lists of descriptors are used to record the results of evaluation experiments and the passport data. Ideally, descriptor lists, containing information on such characteristics as plant height and seed color, are uniform for each species and are designed for rapid entry of data in standardized form into a computer. IBPGR regularly publishes descriptor lists for crops and their wild relatives; recent examples include lists for pear and kodo millet (*Paspalum scrobiculatum*) (IBPGR, 1983d,e). Gene banks that have computers enter coded characteristics into data bases. Information on an accession may be in a file held at a central mainframe computer to which several microcomputers are connected, or in several different files at terminals linked to a mainframe computer. In such cases, the data-base management systems should be designed so that the files can be merged if necessary. At the International Potato Center (CIP—Centro Internacional de la Papa), for example, a request for accessions with certain characteristics is typed into a terminal keyboard, and the computer screens the accession records for the desired characteristics; the printout is then torn off and samples are taken from the appropriate accessions. And, as another example, the gene-bank computer at IRRI can search for thirty-eight morphological and agronomic traits for each accession (Chang, 1980).

Germplasm of seed crops is stored in three main types of collections: short, medium, and long term. In working or short-term collections, seeds are kept at room temperature, or if the climate is hot and humid, in air-conditioned rooms. Materials in working collections are of current interest to breeders and are used at least once a year. Working collections are outside the framework of genetic conservation and are usually kept in rooms assigned to crop-breeding programs, rather than in a central gene bank. Samples from short-term collections are sometimes eventually incorporated into generally larger medium- and long-term collections in an institute's genetic resources conservation unit. Many of the accessions at the gene bank operated by the International Center for Agricultural Research in the Dry Areas (ICARDA) near Aleppo, Syria, for example, originated in this manner.[1]

[1] For information on the founding date and mandated crops of ICARDA and other international agricultural research centers within the Consultative Group on International Agricultural Research, see Appendix 1.

4.2. Dr. T. T. Chang, head of the gene bank at the International Rice Research Institute, Los Baños, Philippines, 1985, with rice accessions vacuum-packed in aluminum cans.

The IBPGR has supported research to help predict the behavior of seeds of many species under reduced moisture and temperature. The real key to successful seed storage is to dry the seed and keep it at a low moisture content; these procedures are frequently more important than temperature conditions. Once the moisture content has been reduced, the lifespan of seeds can be expected to double for every 5° C that the temperature is lowered between 50° and 0° C (Harrington, 1970).

In medium-term storage, where material of potential interest to breeders is stored, seeds are kept close to freezing in glass or plastic bottles or aluminum-foil packets. Breeders regularly tap accessions in this type of storage. Most gene banks are designed for short- and medium-term storage because they serve the immediate needs of breeders, and electricity costs are lower than for long-term collections.

For long-term storage, seeds are dried and sealed in bottles, or vacuum packed in cans (Figure 4.2) or in laminated aluminum-foil envelopes, and placed on shelves in a well-insulated room or freezer where the temperature remains in the −10° to −20° C range; under such frigid conditions seeds can remain viable for extended periods, perhaps for as long as a century. The Nordic gene bank uses glass bottles because of the possibility that plastic may emit mutagenic gases; even though the amount of volatile material may be small, over time the reagents could damage stored germplasm (Yngaard, 1983). Another advantage of glass bottles is that they allow silica gel, which is sometimes packed with samples to keep them dry, to be seen; a color change in the silica gel packets indicates that moisture has entered the sample. On the other hand, glass bottles break more easily and accessions can become mixed if several bottles fall together.

Samples in long-term storage are often referred to as "base collections" and are not normally used for routine distribution or exchange; they are security collections (Hanson et al., 1984:1). Base collections are often similar or identical to material in medium-term collections that provide samples to breeders and other scientists for evaluation. Seeds destined for medium- or long-term storage are gently dried to a moisture content between 4 and 7 percent before freezing. Many gene banks contain an anteroom or buffer area equipped with an air conditioner and dehumidifier to reduce moisture and temperature fluctuations of the air in the seed storage-room. Some institutes, such as IRRI and the maize collection site at CIMMYT, contain facilities for both long- and medium-term storage, while others, such as NSSL at Fort Collins, Colorado, contain only collections under long-term storage.

Once in a gene bank, accessions are periodically checked to make sure they are still viable. Three main factors govern the viability of stored seed: temperature, seed-moisture content, and initial viability. At least 90 percent of seeds in a sample destined for storage should germinate (Roberts, 1983). To test for viability, a subsample is taken from the accession and germinated, often on filter paper. If germination falls below 85 percent, the accession is regenerated to avoid loss of rare genes and because genetic changes begin to occur when seed viability declines. Some gene-bank operators set higher limits, since some genes are extremely rare in a population, and even a slight deterioration of an accession's viability could mean the loss of some potentially valuable genes. A lower germination rate is acceptable for relatively modern varieties because they are genetically more homogeneous than landraces or wild species. Gene-bank operators set regeneration standards for each crop species and must balance the risk of valuable gene loss against the high cost of regeneration.

Germplasm of root crops and of some seed plants (Figure 4.3) is held as a working collection in field plantings or in tissue cultures. If potato germplasm is maintained in tuber form, material must be planted annually and the tubers stored in a cool, dry place (Figure 4.4). Cassava, on the other hand, can remain in the field for several years before stem cuttings are taken and plants propagated. At CIAT (Centro Internacional de Agricultura Tropical) near Cali, Colombia, however, accessions in the cassava field gene bank are replanted yearly to reduce disease and pest damage. Some root crops set seeds that are excellent for

4.3. Field gene bank for several species of *Tripsacum*, wild relatives
of maize used in breeding for resistance to diseases, insect pests, and
resilient stems. Tlaltizapan substation of the International Maize
and Wheat Improvement Center, Mexico, June 1985.

conserving genes, but true seed cannot be used to preserve the genetic
identity of an individual clone.

From the viewpoint of genetic conservation, seed storage of root
crops and other clonal material, such as sugar cane, is ideal. A limited
number of clones would be maintained in a field gene bank or in tissue
culture when they represent landmarks in breeding, are sterile, or are
well known genetically. Breeders naturally need to maintain many
more clones, but they are not always necessary for genetic conservation
in the long term (IBPGR, 1985a). The genes of other clonal crops, such
as temperate fruits and grapevines, could also be conserved as seed.

Because of the relatively large space requirements and high labor
costs of field maintenance, root crop germplasm is increasingly stored
in vitro.[2] At CIP's Huancayo substation in the Peruvian Andes, for example, a 4-hectare field is needed to plant 6,000 potato clones every

[2] Tissue-culture techniques for storing crop germplasm are discussed more fully in Chapters 5 and 6.

year. At CIP's headquarters in Lima, in contrast, facilities have been built to house a duplicate collection in tissue culture that will require only 0.1% as much space (Smith, 1983a). CIAT's 3,700 cassava accessions in the field gene bank occupy 8 hectares. As of August 1985, approximately two-thirds of the collection was also stored *in vitro*, and the entire collection is expected to be duplicated in tissue culture form by 1987 (W. Roca, pers. comm.). The tissue culture gene bank for cassava, which can house 6,000 accessions, is only 7 meters long, 6 meters wide, and 2½ meters high.

The International Institute of Tropical Agriculture (IITA) at Ibadan, Nigeria, also maintains germplasm of tuberous plants in tissue culture and in field plots. Test-tube germplasm is generally kept in windowless rooms equipped with air conditioners and fluorescent lights. The relatively cool temperatures retard growth, and day length can be controlled with artificial light. With current techniques, test-tube cultures of certain crops, such as sweet potato and cassava, can remain healthy for up to two years; thereafter subculturing is needed to regenerate the cultures.

All tissue-culture collections are, however, only temporary germplasm stores. Furthermore, because some materials exhibit genetic instability in tissue culture, safer and long-term storage methods for clonal materials are being sought. To this end, experiments are being conducted for the long-term storage of crop tissue cultures in liquid nitrogen at $-196°$ C (cryopreservation). At the extremely low temperature of liquid nitrogen, animation and genetic change should be suspended.

PROBLEMS AND PREVENTIVE MEASURES

Just as scholars depend on libraries to conduct their work, crop breeders need access to germplasm collections to develop varieties. But difficulties have surfaced in this area, some of which can be overcome with foresight and planning, whereas others require long-term study, training, and education. And some are common to all gene banks, while others occur infrequently.

With regard to gene-banking principles, a few agricultural institutes under national control may not always adhere to the principle of free exchange of germplasm. While gene banks at international centers within the Consultative Group on International Agricultural Research (CGIAR) respect this principle, national gene banks may occasionally refuse to release accessions. Such export embargoes may be imposed by

4.4. Dipping potatoes (*Solanum* spp.) destined for germplasm storage in baths containing fungicides, pesticides, and bactericides at the International Potato Center's substation near Huancayo, Peru, June 1982.

politicians who perceive that it is not in their national interest to supply genetic material to certain countries or organizations.

Thus far, there is no documented evidence for the non-availability of germplasm of the major food crops, and the few cases where germplasm has been withheld all involve export-oriented cash crops. Coffee breeders find it impossible to obtain material from Ethiopia (MacFadyen, 1985), the home of coffee. Wild and cultivated coffees contain varying degrees of resistance to coffee rust (Sylvain, 1955), and coffee breeders, particularly in Latin America, are keen to obtain fresh germplasm with resistance to this widespread disease. The Jamaican government will not allow germplasm of allspice (*Pimenta dioica*) to leave the country in an apparent attempt to monopolize production. Other political decisions to ban the export of crop germplasm include black pepper and turmeric (*Curcuma longa*) in India, cacao in Ecuador, sugar cane in Taiwan, date palm in Iraq, rubber in Brazil, African oil palm in Malaysia, and pistachio (*Pistacia vera*) in Iran (Witt, 1985:103). And following pressure from the U.N. Food and Agriculture Organization (FAO), Sudan lifted an embargo on the export of gum arabic (*Acacia senegal*) germplasm (Prescott-Allen and Prescott-Allen, 1983:69).

In most cases, though, a failure to honor a request for germplasm is not due to political considerations. More often the accession needs to be multiplied before sufficient seeds are available for exchange, because quarantine restrictions apply, or because the accession may be in a base collection from which material is not sent out. Applicants for germplasm samples sometimes write to base collections when such requests should be sent to active gene banks. To help overcome this problem, the IBPGR has issued directories of holdings so that applicants can write to the correct institutions. It should also be remembered that a sample is rarely unique; it can usually be found in several gene banks. For example, wheat accessions in the Small Grains Collection of the U.S. Department of Agriculture, which is the world's largest supplier of material, are also held in full or in part in the Soviet Union, Italy, ICARDA, Brazil, Ethiopia, Turkey, and many other countries.

The potential for periodic restrictions of germplasm exchange was a powerful argument for establishing a network of international and national gene banks that subscribe to the principle that all bona fide researchers of any nationality who can use the material for the benefit of mankind should have access to germplasm. Fortunately, cooperation has historically been the rule of thumb in germplasm exchange because it is difficult to predict where one will have to turn for help to solve a crop-breeding problem. Almost all agricultural research institutions thus abide by the principle of free exchange of germplasm.

Duplicate collections held at different locations help circumvent problems of germplasm access and avoid a total loss of unique germplasm in the event of power failure, fire, or natural hazards. The supply of electricity in Third World countries is often erratic, particularly away from capital cities. A short interruption of power to a gene bank is unlikely to jeopardize collections, but a prolonged outage can lead to spoilage. Wildly fluctuating voltage is fairly common in developing countries and sometimes causes failure of refrigeration equipment. A medium-term gene bank installed in 1978 by the Research Institute for Food Crops at Bogor, Indonesia, for example, never functioned properly because of voltage problems; the facility has therefore been disconnected and is used for storing office supplies.

Servicing of broken-down equipment is often difficult in developing countries. Iran, for example, has lost valuable stored germplasm because of the failure of refrigeration equipment. Bilateral agencies are partly to blame for the failure of equipment when they insist that Third World countries acquire machinery from the donor country regardless of whether maintenance facilities are available in the recipient country. In the case of the long-term gene bank for legumes operated by the National Institute of Biology at Bogor, for example, twenty freezers with

4.5. Easy-to-maintain chest freezers used for long-term storage of barley and durum wheat germplasm at the International Center for Agricultural Research in the Dry Areas near Aleppo, Syria, 1984.

external thermostats and temperature gauges were received from the United Kingdom in 1981. Only half of them worked on arrival, apparently due to damage during shipping. In January 1985, only five from the original shipment were functioning and no repairs had been made. Simple freezers designed for household use can also store seeds adequately under long-term conditions; such freezers are sold in most developing countries and can be serviced locally (Figure 4.5).

82

The expert seed-storage committee of the IBPGR has condemned the profligate waste of money on the buildings and equipment of gene banks when the money could be spent more effectively on more modest facilities and on better-trained scientists to operate them adequately (IBPGR, 1985b). Few gene banks, for example, have heeded warnings that they need a competent seed physiologist on their staff. The IBPGR is currently establishing a register of gene banks that meet preferred scientific standards, which should help fine-tune the world network in meeting minimal international scientific standards. Simply building a gene bank is no longer sufficient grounds for automatic inclusion in the network.

To help overcome the vulnerability of germplasm collections to voltage and equipment problems, voltage regulators, as well as backup cooling equipment and generators, are recommended. A gene bank should contain at least two air-cooling units that can be operated in relay fashion; in this manner maintenance can be performed without endangering the collections. Even with backup equipment in place, improper wiring can ruin germplasm; at the University of Viçosa in Brazil, for example, a short-circuit fire largely destroyed a lima bean (*Phaseolus lunatus*) collection.

Some countries are able to save energy costs by storing germplasm in naturally cold and dry locations. Argentina, for example, is experimenting with the storage of cereal accessions in ice in Antarctica (A. Von der Pahlen, pers. comm.). Since 1984, the Nordic gene bank has used an abandoned mine on Spitsbergen Island to store an initial batch of 180 accessions. Spitsbergen, at 78° N latitude, is well within the Arctic Circle, and samples are stored in a mine shaft 70 meters below the surface (A. Wold, pers. comm.). Samples are thus kept at a constant $-3.7°$ C. Seeds are dried to 5 percent moisture and placed, five hundred per accession, in sealed glass ampules. The accession number is etched on the outside of the ampules, and the samples are sent by air to Spitsbergen inside insulated containers.

Solar-powered gene banks are also on the drawing board and hold promise for some temperate and tropical countries. One passive configuration for cool, dry highland locations, such as parts of the Andes, requires no electricity for maintaining samples at $-4°$ C. Temperatures are kept below freezing inside the well-insulated passive solar gene bank by a mixture of water and alcohol. Ice is generated during winter and at nights through rapid radiation cooling; no fans are needed since air circulates by thermal convection. For temperatures as low as $-15°$ C, a second chamber, equipped with refrigerators powered by photovoltaic cells, is created inside the passive solar unit (Saravia and Lesino, 1983).

Even gene banks with a constant power supply and regularly serviced machinery have some difficulties with maintaining samples because of staff shortages, lack of storage space, and limited facilities for regenerating materials. This can result in only a sporadic checking for seed viability and an infrequent inventory of the stock on hand. Popular accessions deplete rapidly, and the rest can seriously decline in quality and viability and may be in urgent need of regeneration. At the National Seed Storage Laboratory in Fort Collins, Colorado, for example, the viability of some accessions has fallen as low as 60 percent due to the problems mentioned above (Murata et al., 1981).

Even when accessions are checked regularly for viability and sufficient stocks, seeds that are still viable after many years of storage may not always produce vigorous seedlings. Furthermore, unwanted changes can occur during regeneration due to genetic drift (change in the genetic constitution), accidental hybridization, and novel selection pressures during grow-out (Allard, 1970; Frankel and Soulé, 1981:237). To reduce selection pressures while regenerating maize accessions at CIMMYT, samples are grown in a day-length neutral environment, and scientists try to create a safe environment by applying fungicides and pesticides to regeneration plots.

Ideally, accessions requiring regeneration should be returned to their areas of origin for planting. It is easier for a gene bank with relatively small geographical coverage to approach the ideal way of regenerating germplasm. The Hungarian national program, for example, has divided the country into major agro-ecological zones, and cooperating organizations within each zone have been identified to grow out accessions. The Mexican agricultural research program, INIA (Instituto Nacional de Investigaciónes Agricolas), employs a similar strategy when it regenerates samples held in the medium-term gene bank at Chapingo. Accessions at this gene bank, mostly maize and beans, are regenerated at INIA's network of stations throughout the country, which covers a diverse range of climates and soil types. And at nearby CIMMYT, highland maize accessions are regenerated at the El Batán headquarters at 2,240 meters, while lowland maize germplasm is renewed at the Tlaltizapan substation, 940 meters above sea level. To regenerate maize samples sensitive to day length, CIMMYT has recently made arrangements with Ecuador's national program for such accessions to be grown out at Santa Catalina, and a similar accord is being worked out with the Peruvian national program. Other gene banks need to formalize similar cooperative arrangements in order to regenerate material properly.

In the case of outcrossing crops, where individual plants, such as maize, rye, beet, faba bean (*Vicia faba*), and many forage plants interbreed, accessions being planted for renewal should be widely separated in space or time to reduce the chance of pollen straying between accessions. Forage acquisitions have occasionally been planted too close to each other because of inadequate facilities and too few staff (Timothy and Goodman, 1979). A separation of at least 3 kilometers between insect-pollinated accessions is desirable, although sizable physical barriers such as tree stands may serve to reduce the distance somewhat (Oka, 1983). At the Gatersleben gene bank in the German Democratic Republic, insect-pollinated crops are covered by screened cages containing several bumble bees to prevent outcrossing (Lehman, 1979). In the case of maize, a wind-pollinated crop, 300 to 500 meters of separation between accessions being regenerated is generally adequate. Maize farmers in Central America maintain the integrity of their different landraces by staggering the flowering times of their maize crops; a similar approach is sometimes used by gene-bank operators. Another way to maintain the genetic integrity of maize accessions is to half-sib the plants by hand pollination on ears protected from open pollination (one row is the female parent while the other provides pollen), a procedure followed by CIMMYT and Mexico's national program. At research institutes with sizable grounds and staff— such as at ICARDA, which has collections of faba bean— accessions are kept well apart during regeneration.

Mislabeling of accessions can also occur during regeneration, and the more times material is renewed the greater the chances that samples will be incorrectly identified. IRRI is one of the few gene banks that maintains a reference seed file containing packets of the original accessions to help sort out samples that have been incorrectly identified during regeneration (Chang, 1976a:16). A reference sample of each germplasm accession is also maintained in glass vials for squashes (*Cucurbita* spp.), peppers (*Capsicum* spp.), and annatto (*Bixa orellana*) at CATIE (Centro Agronomico Tropical de Investigación y Enseñanza), a regional agricultural research and germplasm center on the outskirts of Turrialba, Costa Rica.

Wild species pose additional problems for curators of germplasm collections. Seeds of cereal-crop relatives and wild legumes tend to shatter when ripe, thus spilling on the ground. Therefore, one of the first characteristics selected when cereal and grain legumes were domesticated was a nonshattering seed head or pod. Wild species need to scatter seeds for survival; domesticated plants have usually lost that ability and depend on man for propagation. Seed dormancy, a survival mechanism for wild species, is another problem sometimes encountered with germplasm of crop relatives. Unless special tests are performed, a dormant seed may appear to be nonviable. Humans have selected against dormancy during plant domestication to ensure uniform crop stands with good yields.

Data about germplasm accessions are nearly as valuable as the plant material itself. Information, whether on paper or computer tapes or disks, is vulnerable to fire and flooding, so it is necessary to keep duplicate sets of accession records in other locations. Outdated formats of gene-bank records are often another problem, with information at some gene banks still being kept on cumbersome file cards or in log books, a practice that not only slows information retrieval but requires the use of copying machines to make duplicate records. Sometimes the copies are not even legible, particularly in the case of penciled field notes.

Computerized gene banks are not without their problems, however. One runs into difficulties when trying to forge a network of computerized gene banks or to compile a master list of all accessions of a crop. Gene banks use many different computer brands and software programs, and they are often incompatible. In a recent survey of fourteen gene banks in Europe and the Middle East, for example, eight different makes of mainframes and four different makes of microcomputers were in use (UNDP/IBPGR, 1984:19). The number of different models was even higher, and not even all of those models made by the same manufacturer are fully compatible.

The results of tests designed to evaluate the performance of samples under environmental stresses, such as disease and insect attack, are of utmost importance for accession records. But evaluation of accessions is sometimes neglected or postponed because it is one of the most costly and time-consuming aspects of gene banking. At some gene banks, virtually all the accessions have been evaluated. At ICRISAT, for example, 84 percent of sorghum, 89 percent of groundnut, 94 percent of chickpea, 96 percent of pearl millet, and virtually all pigeonpea accessions have been evaluated (Mengesha, 1984). Most of CIAT's thirty-seven hundred cassava accessions have been evaluated in six edaphoclimatic (soil and climate) zones, five of which were in ecologically diverse Colombia. At other gene banks, however, only limited passport data are available. Of the world's germplasm collections, some 65 percent have no passport information and between 80 and 95 percent lack characterization or evaluation data (TAC, 1985:37). Even at well-managed gene banks, many potentially useful genes remain unused because evaluation has lagged behind the pace of collection. Three-quarters of IRRI's rice collection has been evaluated, but only 10 percent of CIP's potato accessions have been thoroughly tested for their characteristics (IRRI, 1984:9; CIP, 1984:6).

A major reason for inadequate evaluation is that it requires a team of highly trained specialists—a staff capability found only at few gene banks and associated agricultural research institutions (Wilkes, 1984). Pathologists, entomologists, physiologists, and agronomists are needed to perform an initial evaluation, which usually entails observations of planted accessions in glasshouses and field plots. Another reason for the lag in evaluating accessions is that breeders are usually busy with materials of current interest in the preparation of new varieties; this undoubtedly accounts for the fact that less than half of the maize collection maintained by CIMMYT in Mexico has been evaluated. Therefore, rather than having to rely on plant breeders who are occupied with other tasks, institutions need special staff to evaluate accessions quickly.

Another hindrance to regenerating and evaluating accessions adequately is the high degree of redundancy in some collections (IBPGR, 1984a:2). Duplicate collections are necessary, but excessive exchanges lead to redundant material that occupies valuable space and slows regeneration and evaluation work (Holden, 1984). Electrophoresis, which is essentially a way of protein-fingerprinting accessions, can help differentiate plant germplasm. With this method, genetic variability can be assessed and accessions can be separated accordingly (Esquinas-Alcazar, 1981:15; N. Smith, 1983a). CIP has used electrophoresis to trim the number of accessions in its potato collection from 13,000 to 6,500 (CIP, 1984:58).

Plant quarantine procedures sometimes interrupt the smooth functioning of a gene bank. Usually, plant material is inspected for potential pathogens or insect pests and promptly released; however, delays are occasionally experienced, thereby jeopardizing materials. Also, governments may impose temporary or permanent bans on the importation of certain plant material in order to keep out disease and insect pests. CIAT, for example, maintains a base collection of *Phaseolus* beans to meet its global responsibilities, but Colombian quarantine officials are strict on bean germplasm imports, particularly from Africa. And near Hyderabad, India, the International Crops Research Institute for the Semi-Arid Tropics (ICRISAT) runs a long-term gene bank with world coverage for sorghum, among other crops; unfortunately, the sorghum collection contains few wild or weedy species, partly because of stringent quarantine measures imposed by the Indian government.

ICRISAT operates a base collection for groundnut but has problems acquiring wild groundnut material and vegetative clones because of quarantine restrictions.

In order to reduce the chances of spreading pests and pathogens, care must be taken when shipping accessions. Quarantine officers hold up accessions if extraneous material, such as soil or foreign plant matter, is evident (IBPGR, 1983a:4). Gene banks, particularly at international centers, thoroughly check incoming and outgoing materials for insects, nematodes, bacteria, fungi, and viruses. One of the worst nightmares for a gene-bank operator is to be accused of letting a destructive disease or pest slip into a country. At CIP, genetic resource specialists scrutinize any material that is bound for export, and in collaboration with Peruvian authorities, they issue a phytosanitary statement that certifies that sanitation standards exceed those required for commercial quarantine acceptance by importing countries (CIP, 1984:6). CIP uses a number of tests to check for disease in germplasm samples destined for shipment, and routinely screens the samples for more than twenty different viruses. One serological method uses antibody-sensitized latex particles while another involves enzyme-linked immunosorbent assays (ELISA). Although these tests are high-tech, they are not necessarily expensive. At this printing, the ELISA kit costs only $250, and DNA probes are becoming cheaper. Despite precautions, however, customs agents sometimes detain material destined for gene banks. Passage of valuable material through quarantine checks can be expedited if importing institutions have secured the necessary clearances. Once the material gets to them, many gene banks grow it in glasshouses or screenhouses to check for signs of disease before it is incorporated into their collections.

Questions inevitably arise about the competency of plant quarantine officers charged with issuing phytosanitary certificates for export and import. In the case of potato, for example, 266 pests and pathogens are known to attack the crop, which is grown in over one hundred nations. Just keeping track of the distribution and evolutionary changes of potato diseases and pests is a monumental task. Clearly the issue of plant quarantine procedures warrants further study. At the very least, more well-trained plant quarantine officers are needed who have access to the latest information on pest outbreaks.

PLANTS WITH RECALCITRANT SEEDS

A number of crops produce seed that cannot be dried or stored at low temperatures and therefore pose special problems for germplasm stor-

4.6. Durian (*Durio zibethinus*) for sale in
Bogor, Java, January 1985. The sticky,
cream-colored pulp of durian is relished
in Southeast Asia.

age. Such crops include rubber, cacao, palms, numerous tropical forest
species, and many tropical fruits (Withers, 1980:1; Van der Maesen,
1984; R. D. Smith, 1984). Durian (*Durio zibethinus*; Figure 4.6), for example, is a delicacy in Southeast Asia, but its seeds cannot withstand
drying or freezing. The creamy pulp surrounding large, oval durian
seeds has a strong cheeselike flavor and the fruit emits a powerful odor,
presumably to attract animal dispersal agents in its ancestral jungle
home.

Germplasm of species with recalcitrant seeds can be maintained in a
number of ways. The most common method is storage in *ex situ* field
gene banks. However, such plantings are damaged by pests, diseases,
and storms and suffer from high labor costs. If cryopreservation techniques are perfected and tissue-culture methods are worked out for
plants warranting germplasm storage, *in vitro* gene banks will play an
important role in the future. Wild species should also be safeguarded in
in situ preserves.

89

4.7. Guava germplasm collection, Cibinong Garden, near
Bogor, Indonesia, 1985.

Botanic gardens, although not currently significant in germplasm
storage on a global scale, could play a larger role as *ex situ* gene banks
for species that have recalcitrant seeds (N. Smith, 1986). Several botanic
gardens have already taken the initiative in conserving genetic diversity
of certain tree crops. The Calcutta Botanic Garden, for example, maintains a germplasm collection of *Citrus* (Sharma, 1984). Bogor's 40-hec-
tare satellite garden at Cibinong maintains a small collection of local banana (*Musa* spp.) clones, guava (*Psidium guajava*; Figure 4.7), and
Annona spp. Cibinong, opened in 1980, is projected to grow to 150 hectares to accommodate larger fruit germplasm collections. The 47-hec-
tare Kahanu Garden on Maui, part of a network of Hawaiian gardens
operated by the Pacific Tropical Botanical Garden, maintains forty-five
cultivars of breadfruit and a smaller collection of coconut (Theobald,
1982). The headquarters of the Pacific Tropical Botanical Garden at
Lawai, Kauai, also has collections of fruit trees important in subsistence
and commerce in many tropical countries; the lush and sinuous 177-
hectare garden has approximately one hundred banana cultivars and
five hundred palm species. The 48-hectare Lyon Arboretum of the

University of Hawaii, Honolulu, maintains two hundred accessions of taro, with eight plants per accession.

Botanic gardens with satellite gardens covering a range of environments are particularly good candidates for germplasm conservation for economic plants with recalcitrant seeds. Bogor, for example, has five daughter gardens covering 866 hectares in a diverse array of climates and soil types, while the Foster Garden in Honolulu has four associated gardens on Oahu that occupy 315 hectares (Sastrapradja and Davis, 1983, 1984; Sastrapradja et al., n.d.; Weissich, 1982). For botanic gardens with a single site, germplasm work would be furthered if they formed research networks. India, for example, has fifty-five botanic gardens operated by various government agencies, universities, and autonomous organizations that would profit from coordinating their programs (Sharma, 1984). A network of botanic gardens could divide the responsibility for conserving certain species according to the climatic conditions and research capability of each.

Botanic gardens also advance the cause of preserving plant germplasm by advising conservation programs about priority sites for safeguarding natural habitats. The staff of the Bogor Botanic Garden, for example, assists the Indonesian government in the selection of nature reserves (Sastrapradja, 1982). Staff at botanic gardens, perhaps in collaboration with scientists at other institutions, can help elucidate the complex interactions between species in tropical environments, particularly pollinating mechanisms of fruit and timber trees. Such information will prove vital for storage efforts and future breeding work.

The herbaria associated with many botanic gardens can also provide a valuable backstopping service in the effort to conserve and utilize wild species by helping to identify and locate crop relatives. The herbarium of the Singapore Botanic Gardens, for example, contains over 600,000 specimens. Over three decades ago, Vavilov (1949) emphasized the importance of systematics for plant breeding. More botanical surveys are needed in the tropics to identify species and map their occurrence (Raven, 1976). Armed with a clearer idea of species distributions and their variability, scientists will be able to gain a better idea of the size of gene pools and where germplasm collections need to be made.

Hawkes (1983:127) suggests maintaining some recalcitrant seed species as slow-growing seedlings under low light conditions. This technique could be used for tropical forest species that are accustomed to shaded conditions, particularly during the early growth stages. In jungles, some tree seedlings grow extremely slowly until a light gap opens, permitting more rapid development. Follow-up research on this idea is warranted.

More research is also needed on *in vitro* cultures. Emphasis in tissue culture research has thus far concentrated on mass propagation and on detecting viruses and viroids in cultures. Insufficient research has been directed towards slow growth and maintenance of stability during storage. Short-term storage of potato and cassava in tissue culture is reliable, but such techniques have yet to be perfected for plants with recalcitrant seeds.

For crops with recalcitrant seeds it is necessary to find practical ways to conserve sufficient accessions that represent the genetic variability of gene pools. The gene pools of cacao and rubber, for example, cover vast tracts of Amazonia, and in the case of cacao, reach as far north as Mexico. Germplasm conservation thus requires a number of complementary storage methods. For some crops, *in situ* conservation is thus necessary to complement field gene banks.

THE *IN SITU/EX SITU* CONTROVERSY

Given the potential hazards associated with the housing of germplasm in gene banks, some scientists feel strongly that the genetic diversity of crops and their wild relatives is best preserved *in situ*, that is in farmers' fields and in natural habitats (N. J. Brown, 1982; Myers, 1983:25). Gene banks, it is argued, arrest the evolution of plants and thereby preclude the possibility of developing new species and varieties; *ex situ* germplasm collections have even been described as "genetic ghettos" (Myers, 1983:123). It is true that frozen seeds no longer interact with their environment, and may even undergo unexpected and unwanted changes during regeneration, but they are nevertheless valuable resources conveniently accessible to breeders.

An analogy can be drawn here with zoos. Zoological parks, like gene banks, are artificial environments where a part of the natural repertoire of animal behavior and other characteristics is inevitably lost. But unlike zoos, gene banks attempt to secure a large part of the genetic variability of a species, something that is difficult to do with animals in artificial enclosures. Some zoos, such as the one at San Diego, California, are maintaining frozen sperm of certain wild animals in order to save genes that may be lost from threatened populations in the wild. Frozen semen and embryos are thus one strategy for safeguarding at least some of the genetic diversity of wild animals. Gene banks perform a similar function for plant genetic resources.

The preservation of landraces in places where they originated has considerable merit. In this manner they would be maintained under conditions to which they are adapted and could continue to evolve.

However, a number of practical considerations arise. Whether land-races are maintained *in situ* or in gene banks, their survival depends entirely upon people. Proponents of *in situ* conservation of genetic resources do not spell out in detail how the thousands of landraces of crops would be tended. Various writers have suggested that farmers be offered subsidies to maintain old varieties threatened with abandonment (Myers, 1983:25). But how would such a program be administered? The task of monitoring farmers who would participate in *in situ* conservation programs is daunting, especially in isolated regions. Extension staff are either nonexistent or already overworked and in short supply, and the potential for corruption is enormous (Frankel, 1970). Also, a sure recipe for alienation exists when extension workers are asked to police farmer activities. It is not at all clear either how much of the subsidies would end up in the pockets of farmers maintaining old landraces. No one has the moral right to coerce farmers to grow low-yielding landraces while others are adopting high-yielding cultivars (Hawkes, 1977a). It is also uncertain how much care "museum" landraces would receive from cultivators who would not otherwise plant them. Weeds and insects could proliferate and severely damage old landraces, even under subsidy, while farmers tend their preferred crops.

Another serious drawback to the idea of the conservation of old landraces in farmers' fields is the scarcity of arable land in many countries (Arnold et al., 1986). In Bangladesh, for example, 100 million people are squeezed into an area the size of Wisconsin. The average size of a Bangladesh farm is under one hectare. It is not clear how farmers in such countries could afford to set aside land for maintaining crops not destined for domestic consumption or for sale. The subsidy for continuing the planting of unused landraces would have to be sizable to compete with subsistence and cash crops; few Third World countries could afford such a program. Myers (1983:24) proposes levying a tax on commercial seed companies to underwrite the cost of maintaining the germplasm of landraces and wild species in farmers' fields and in natural preserves. It is doubtful whether such a measure is operationally possible, and it leaves unanswered the question as to why publicly owned seed operations should not also be called upon to support germplasm conservation.

The idea that landraces could be grown on plots maintained by national programs in areas where they were formerly cultivated has also been floated. Again, costs and priorities undercut this proposal. Land would have to be purchased or leased for such a program and scarce staff diverted from agricultural institutes to monitor these vignettes of the past. Besides, *in situ* conservation is not foolproof; landraces and wild species can disappear outside of gene-bank buildings as well as within (Prescott-Allen and Prescott-Allen, 1982a).

Gene banks have saved landraces and populations of wild species from extinction in several instances. An accession of *Oryza perennis* from Taiwan that was found to resist ragged stunt virus is now extinct there; fortunately, collections of the cosmopolitan species were made in Taiwan and deposited in IRRI's gene bank before the island strain disappeared. In 1971, following a severe earthquake, Nicaragua lost its entire germplasm collection. The same year, CIMMYT sent eighty-four maize accessions that originated in Nicaragua to help replenish the nation's destroyed gene bank. In 1975, extensive damage occurred to Nicaragua's reassembled germplasm collections due to interruptions in the supply of electricity; CIMMYT responded positively to a request for fifty-two accessions. And in 1977, CIMMYT sent to Managua an additional twenty-eight accessions of maize from Nicaragua.

In Kampuchea, many unique rice varieties were lost in the 1970s when war disrupted agricultural production. Seeds of numerous landraces were eaten or rotted, so the distinct varieties died out. Luckily, the gene bank at IRRI contains rice varieties that were collected in Kampuchea before the outbreak of political strife, and some of these have been successfully reintroduced to the country. In 1981, for example, IRRI sent 36 Khmer varieties to Kampuchea through the offices of OXFAM, and an additional 103 indigenous varieties were received by the national program in 1982. These reintroductions have been multiplied locally for distribution to Kampuchean farmers (Swaminathan, 1984a). Some of the wild barleys collected by Vavilov in Ethiopia and deposited in the Vavilov Institute gene bank in Leningrad became extinct in Africa, though these species have been reintroduced to Ethiopia (Evans, 1975).

Perhaps the most serious drawback of *in situ* conservation of crop germplasm as the sole method of preserving genetic diversity of cultigens is that the material would not be readily available to breeders. If germplasm were only available in farmers' fields, breeders would have to organize costly expeditions each time they needed fresh material. Also, unless regular surveys were conducted, those landraces would remain essentially unevaluated and useless for breeding. The notion of duplicating gene-bank collections in farmers' fields may sound attractive, but funding for gene banks is already inadequate. Farmers always have the option of maintaining old varieties because of their special attributes, such as preferred taste, at their own expense. But if scarce financial resources are to be stretched to maintain *ex situ* gene banks as well as field collections of landraces, the capability of the former would surely suffer. Keeping abandoned landraces alive involves a higher order of magnitude than, say, breeds of cattle, pigs, or sheep. For example, there may be over 100,000 traditional varieties of rice (Chang, 1984a); livestock breeds do not even approach a fraction of that figure.

In situ conservation of wild species, on the other hand, is warranted and much more feasible (Frankel, 1977; Swaminathan, 1983:8; Ingram and Williams, 1984). Wild species should be maintained in natural habitats so that they can continue to interact with their environments. Tropical forests merit special attention for conservation since many important crops originated in jungle environments, and tropical forests are treasure troves for plant genes (Myers, 1984). Administering germplasm of wild species *in situ* is not nearly as cumbersome as a subsidy program for landraces. Once important centers of species richness of crop relatives have been identified and demarcated, regular patrols should suffice to maintain their integrity.

For crops such as rubber, cacao, and tropical fruits, *in situ* reserves and field gene banks are currently the most practical way of conserving large gene pools. Since 1983, IBPGR has funded a study of wild mango species in Borneo that is designed to map the distribution of mango relatives and to identify populations that are conserved in parks. And in Sweden, the Nordic gene bank is conducting modeling studies on *in situ* reserves for forage germplasm (UNDP/IBPGR, 1984:7).

Several governments have already taken the initiative in establishing natural reserves for maintaining plant genetic resources. Ethiopia, for example, is setting aside reserves for coffee germplasm (IBPGR, 1985d:32). And British Columbia, after passing the Ecological Reserves Act in 1971, established several *in situ* preserves for plant germplasm, particularly for timber species. The 86-hectare Chilliwack reserve, for example, contains exceptional specimens of red cedar (*Thuja plicata*), grand fir (*Abeis grandis*), silver fir (*A. amabilis*), sitka spruce (*Picea sitchensis*), and Engelman spruce (*P. engelmannii*); and the 263-hectare Takla Lake reserve contains the most northerly known occurrence of Douglas fir (*Pseudotsuga menziesii*), a potentially important source of frost resistance for this commercially important timber species. Furthermore, the 188-hectare reserve near Clinton, British Columbia, contains germplasm important for the timber and livestock industries. One of the most northerly populations of Ponderosa pine (*Pinus ponderosa*), another significant timber species, occurs in the reserve, as do several dryland forage species (Foster, 1984).

Conservationists concerned with preserving the genetic diversity of crops and their wild relatives have unfortunately come on the scene rather late with regard to the establishment of parks and nature reserves. They increasingly realize that governments are more likely to set aside areas for noncommercial use when genetic resources of crops and their relatives are being safeguarded. Plant genetic resources could, in some cases, provide the clinching argument for setting aside a reserve that might otherwise be mined, settled, or covered by a reservoir. Close relatives of crops could be a vanguard for conserving some natural areas, just as pandas help generate funding and support for a large array of lesser-known, or even unkown, animals. In other words, crop genetic conservation adds an important dimension to the conservation movement.

The designation of many reserves now includes genetic conservation as a rationale, but most of them don't even have a list of plant species, let alone an accurate inventory, maps of major species, or an idea of the diversity represented. Variabilities between populations have not been looked at adequately, and the scientific background for establishing reserves for genetic conservation of crop relatives is poor (Ingram and Williams, 1984; IBPGR, 1985c). The International Union for Conservation of Nature and Natural Resources and the World Wildlife Fund now recognize these problems, and we can look forward to a spurt of scientific activity based on prolonged and detailed fieldwork. But it will take at least a decade before we see a major impact on the conservation and monitoring of crop relatives, gene pools of forest species, and such crops as tropical fruits and rubber.

Clearly the *in situ/ex situ* controversy should not be seen as an either/ or proposition. Debate on the issue will be on a sounder footing when it is on firmer scientific ground and is placed in historical and geographical perspective. We should remember that landraces have been abandoned since the dawn of agriculture. Farmers are fully capable of making rational decisions; they stop growing a cultivar when a better cultivar becomes available. They are not coerced into adopting modern varieties. Rather than try to revive abandoned landraces, their seeds and clones should be safeguarded in gene banks so that they provide a genetic reservoir for breeders, now and in the future (Frankel and Soulé, 1981:182; Prescott-Allen and Prescott-Allen, 1982b:81; Swaminathan, 1983:8). Also, many regions of the Third World have yet to experience modern farming. Sizable areas in developing countries are still planted to traditional varieties. Many crops important for subsistence in the Third World, such as cassava and the various millets, have barely been affected by scientific breeding programs. The genetic heritage of our crops should be collected and safeguarded in gene banks before it vanishes. Gene banks may be the only hope for many wild species as well; in spite of efforts to set up parks in developing countries, many are overrun by squatters and carved up by development projects.

BIOTECHNOLOGY AND GENETIC RESOURCES

With the possible exception of atomic fission or fusion, few scientific topics have recently caught the public's imagination or stirred such excitement as biotechnology. No doubt part of this attention is due to the rush of venture capital into biotechnology firms, but there is also heightened anticipation that biotechnology will provide us with astounding new products and medical and nutritional advances that will significantly improve daily life.

Biotechnology has been defined as "the use of living organisms or their components in industrial processes" (NASULGC, 1983). A somewhat more precise definition is provided by Hardy (1984): "The use of a biological system to produce a product, the use of a biological system as a product or the use of the techniques of biotechnology to indirectly provide a product, process, or service." Broadly speaking, biotechnology includes ideas and methods derived from molecular biology and cell biology, using a host of organisms from microorganisms to man and involving fields as diverse as food science and medicine. It is wrong to equate biotechnology with genetic engineering, which is only one subfield of biotechnology, but one in which dramatic advances in microbiology and medicine have been made by altering bacteria and yeasts through the transfer of DNA. These microorganisms have been transformed into mini-factories to produce insulin, interferons, and other pharmaceuticals.

Many believe that advances in biotechnology will play an important role in crop improvement and in conservation of genetic resources. We agree with this assessment, and in this chapter we discuss some of the strengths and potential problems in applying biotechnology to germplasm conservation and utilization. We will emphasize the need for a partnership between biotechnologists, plant breeders, and germplasm specialists to facilitate continued crop improvement for the benefit of mankind.

Two very important advances with agricultural implications have been made through genetic engineering: the cloning of a gene for a protein of foot-and-mouth disease, the first step in producing a vaccine, and the cloning of a DNA copy of the potato spindle tuber viroid (PSTV) in *E. coli*. By using a DNA probe, potato tissue can now be

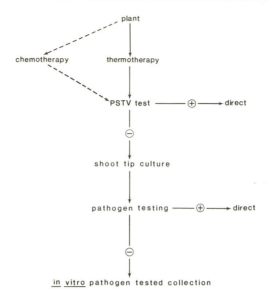

5.1. Cleaning-up procedure for potato material destined for germplasm storage or exchange at the International Potato Center, Lima, Peru. Plus signifies pathogen detected, minus no pathogen detected.

screened for the pathogen. The International Potato Center (CIP— Centro Internacional de la Papa) uses this DNA probe to eliminate material destined for germplasm storage or exchange (Figure 5.1).

BIOTECHNOLOGY AND PLANT BREEDING

Plant breeding has made use of research developments in cell biology for years. Applications such as cell and tissue culture have proved useful in breeding a number of crops, a subject to which we will return later in the chapter. For now we restrict ourselves to a discussion of the molecular biology aspects of biotechnology that are commonly covered under the term "genetic engineering."

Genetic engineering can operate at three basic levels of complexity in plants: (1) introduction of a foreign gene that is expressed at all times and in all organs; (2) introduction of controlling genes that function at specific stages or times of development; and (3) insertion of genes for multiple traits. Most genetic engineering work today is confined to the first level; knowledge is too limited as yet to handle the second and third levels (Bogorad, 1983). Much basic research will be required before the latter two stages can be implemented, and it is there that the greatest advances are expected.

In higher plants, the transfer of single genes conferring resistance to specific diseases or pests, or tolerance to environmental stresses, could have an impact in crop breeding within five to ten years. Overall, though, the shifting of individual genes into crops is not likely to boost yield or quality dramatically for very long. The effect of single-gene resistance is usually ephemeral because the dynamic nature of pests and pathogens typically produces new strategies for overcoming single-gene resistance.

Successful biotechnology must be able to identify, isolate and shift specific genes, as well as ensure their appropriate expression in the target crop. Our present knowledge of what specific genes do and where they are located is still too poor to make it possible for us to find and move such desirable genes. The task of mapping the genes of crops is formidable; each crop plant contains between one million and ten million genes (Shebeski, 1983). The best-studied crops in terms of gene mapping are pea and maize, and only about 170 genes can be accurately located in peas. By comparison, in 1935 more than 300 maize genes were known and about 60 could be identified on chromosomes (Phillips, 1984). Detailed chromosome mapping is needed for each crop and its wild relatives, and even for unrelated plants, if genes are to be transferred effectively. Such chromosome mapping requires a great deal of research before even a small part of the work can be carried out.

Even if the location and action of specific genes are established precisely through more suitable probes, transfer techniques are too crude to be effective in most crops. Plasmids, tiny packages of extra-nuclear genes in the cytoplasm of cells, are currently the most commonly used vectors for gene transfer, though they can be used only with dicotyledonous plants, or dicots.[1] In most cases they have not been successfully used with monocotyledonous plants, a serious drawback considering that the world's major food crops—wheat, rice, maize, sorghum, and barley—are monocots.

Recent studies in the Netherlands indicate that the Ti plasmid of a gall-forming bacterium, a plasmid that is commonly used as a vector with dicots, can transfer genetic information in members of the amaryllis and lily families, which are both monocots (Hooykaas-Van Slog-

[1] Monocotyledons and dicotyledons are the two great divisions of the seed plant world, and many anatomical and morphological differences separate them. Their names derive from the fact that dicots produce two leaves when seeds sprout, while monocots, such as grasses and palms, send out only one leaf upon germination.

teren et al., 1984). While this is a significant development, it remains to be seen whether Ti plasmids, or other vectors, can accomplish the same feat with grasses. Direct gene transfer has been achieved with a grass without the use of a vector— a promising development (M. G. K. Jones, 1985). But much work remains to be done on establishing systems for the isolation, culture, and regeneration of plants from protoplasts; we are still some distance from obtaining a tranformed plant from the "soup" contained within cell membranes.

Scientists at the International Maize and Wheat Improvement Center (CIMMYT—Centro Internacional de Mejoramiento de Maiz y Trigo) are monitoring University of Illinois experiments in which maize pollen is soaked in a solution containing DNA from wild relatives (*Tripsacum* spp.) before the DNA-coated pollen is placed on the silks of maize. Even if fertilization occurs, however, it is still uncertain at this stage whether any foreign DNA is incorporated into the embryo. Australian scientists are also experimenting with pollen tube transfer of foreign genes (Peacock, 1984).

We do not yet have reliable means to determine whether foreign DNA will be incorporated into the genome of a plant and be usefully expressed. To test for inclusion and expression will require a close relationship between biotechnologists who have the necessary skills and experience in molecular and cell biology, and plant breeders and crop scientists who are experts in whole plant manipulation and field evaluation.

Even if a foreign gene can function in a crop plant, its expression needs to be controlled so that desirable characteristics are apparent at the right time. For example, nif genes, which are involved in nitrogen fixation, have been transferred to certain bacteria, but they function sporadically in some bacteria and not at all in others (Hardy, 1983). Barbara McClintock received a Nobel Prize in medicine for her work on gene expression in maize, but we still have much to learn about what switches genes on and off (Swaminathan, 1984b).

More research needs to be done on gene expression mechanisms before genetic engineering will lead to significant payoffs in crop breeding. The most important single attribute of a crop plant, an economic yield of a useful product, is usually governed by a number of genes operating together; drought resistance in wheat, for example, depends on a number of interacting genes, as does tolerance to salinity in crops (Hawkes, 1983:95). In the case of nitrogen fixation in legumes, seventeen nif genes are involved in the symbiotic relationship between *Rhizobium* bacteria and the roots of the host plant. The idea that cereals will

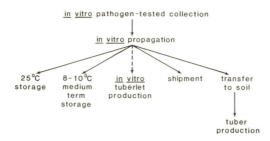

5.2. Use of the pathogen-tested potato collection at the International Potato Center, Lima, Peru.

be self-fertilizing with nitrogen through the addition of nif genes is attractive but unlikely to be realized in the near future.

Many advances in biotechnology still depend on trial-and-error experiments for each crop. This empirical approach will become less important as more information is gained on specific life processes, but for some time it will be necessary to work out techniques for handling and manipulating plants on a crop-by-crop basis.

CELL AND TISSUE CULTURE

Whereas DNA transfers to higher plants by genetic engineering are difficult to carry out and verify, at least until more basic research is done, tissue and cell culture techniques, commonly called *in vitro* cultures, are already useful tools in agriculture. Rapid propagation and multiplication techniques using plant parts such as callus (undifferentiated masses of cells in test tubes or flasks that can be induced to grow plantlets), or shoot tips and buds that can also be cultured to obtain plants for field planting, are being profitably employed in crop breeding. Shoot-tip and bud culture are also used for germplasm exchange and for short-term conservation. CIP, for example, routinely uses tissue culture to prepare and distribute pathogen-tested potato germplasm for storage or shipment (Figure 5.2).

Tissue culture is also used by such institutions as the Hawaiian Sugar Planters' Association, CIMMYT, and other organizations to rescue embryos from widecrosses that might otherwise perish. Anther or pollen culture is employed to raise haploid or double haploid plants that contain only the male set of chromosomes. Haploid plants can be used to produce homozygous plants, which means that they are uniform in genetic characteristics, and it is easier to use such plants to identify undesirable recessive traits and to fix desirable characters (Schroder and

101

Schell, 1983). Haploid plants can speed up the breeding process because doubled haploids are 100 percent homozygous, that is, all of the progeny from these plants share the same traits in one generation, in contrast to the time-consuming selfing or backcrossing usually required in conventional plant breeding. The Chinese have used haploid breeding to develop new varieties of rice, wheat, and maize (Swaminathan, 1984b). The International Rice Research Institute (IRRI) and CIAT (Centro Internacional de Agricultura Tropical) are also using anther culture in haploid breeding; in some cases, though, problems have been encountered in trying to generate whole plants from haploid callus tissue (Khush, in press).

Scientists use cell cultures to identify cells that tolerate high levels of salts, herbicides, or toxins exuded by pathogens and pests. Cell cultures are also used for rapid screening of germplasm for potential tolerance to such stresses as salinity and herbicides. Somaclonal variation—natural, or as some think, induced variation that occurs when cells grown in tissue or cell culture differentiate into whole plants—is being explored as a possible source of new variation for plant breeding. At IRRI, for example, useful rice mutants obtained through somaclonal variation have been selected from tall, traditional salt-tolerant varieties (Khush, in press). CIAT scientists are studying cassava lines that have mutated *in vitro* for possibly useful traits.

In protoplast fusion, another active area of biotechnology research, the cell contents of two plants, related or belonging to different species, are merged. To accomplish this union, cell membranes are chemically ruptured without harming their contents. This technique has allowed scientists in Japan to transform some medicinal plants. Whereas fusion of widely differing plants has been achieved in cell culture, generation of whole, transformed organisms has proved difficult in most cases. Also, le s control over transfer of desirable genes is obtained with protoplast fusion than by using vectors; the technique has yet to produce useful results in major crops, such as the cereals and grain legumes (Burgess, 1984). The technique is something akin to the clumsy shuffling of two packs of cards in which some cards are left out of the new deck while the incorporated ones are in a random order. Unless more control can be achieved, protoplast fusion is more likely to produce bizarre oddities than valuable variation.

INDUCED MUTATION

Induced mutation by chemical means or radiation is not as new a technology as genetic engineering, but can be considered a form of biotechnology because its aim is to transform living organisms. Gamma-ray treatment of seeds has been tried with some success in a number of crops for several decades. Whereas not all mutants have proved useful, the list of plant cultivars derived from induced mutations is impressive (Konzak, 1984; Konzak et al., 1984; Miche et al., 1985). Progress in improving crops has been achieved by radiation and mutation breeding on several fronts, including disease resistance, earlier maturity, tolerance to problem soils, and improved crop quality.

In rice, benefits from induced mutation have included traits that are simply inherited and usually controlled by single recessive genes, notably semidwarfism, early maturity, waxy endosperm, and genetic male sterility (Rutger, 1983). Induced mutation has recently become important in developing parents useful in hybridization programs. By 1982, forty-five rice cultivars had been developed either by direct radiation or by crossing with induced mutants. In Indonesia, gamma-ray treatment of rice-breeding lines derived from Pelita I led to the release of a high-yielding cultivar, Atomita 2, that resists biotypes 1 and 3 of brown planthopper, tolerates salty soils and has good eating qualities (IRRI, 1984:47).

Radiation has also been used successfully to break chromosomes into segments that can subsequently be reattached to other chromosomes. In the case of leaf rust resistance, for example, the responsible gene has been transferred from a wild grass (*Aegilops umbellulata*) to bread wheat (*Triticum aestivum*) (Shebeski, 1983). Chemicals, such as MNU (N-methyl-N-nitrosourea), have also been used to induce mutations and chromosome breakages, but payoffs have so far been modest.

GERMPLASM PRESERVATION

Tissue culture has already yielded benefits in germplasm preservation, particularly for vegetatively propagated crops. This development is especially important for the numerous tropical crops, such as cassava, sweet potato, and yams (*Dioscorea* spp.), that are planted as cuttings. Thus, in the absence of alternative measures, field plantings are required to preserve germplasm. For most vegetatively propagated annual crops, time-consuming and costly replanting is required every year to maintain germplasm and to keep it relatively free of diseases.

Tissue culture preservation is being used or tested on several crops with some success. Clones of cassava, sweet potato, potato, and coffee have been stored as test-tube plantlets for up to two years (Kartha et al., 1981). Most test-tube storage of germplasm depends on slow or restricted growth of the cultured plant sample. Slow growth storage can involve manipulating the test-tube environment by inducing osmotic stress in the growth medium, supplying nutrients in suboptimal or excessive amounts, employing growth retardants, reducing temperature and light intensity, or increasing periods of darkness (Wetter, 1984). Such techniques can be used alone or in various combinations. Most tissue cultures are maintained in the 23-25° C range; slow growth storage can be at temperatures as low as 5-15° C. At CIAT, cassava germplasm in tissue culture storage is kept at 25° C during the day under 500 lux of fluorescent lighting and 15° C during the night. Under such controlled conditions, cassava clones can be stored for 18 to 24 months.

One worry with using tissue culture for germplasm preservation is the occurrence of somaclonal variation in callus and cell-suspension cultures as well as in organized tissues such as shoot tips (Walbot and Cullis, 1985). Such variation is a possible boon for breeders but a nightmare for germplasm specialists who want to avoid genetic change in test-tube storage (Bajaj, 1979; Withers, 1980; Scowcroft, 1984). The amount of induced genetic variation in potato varies depending on the *in vitro* culture system (Figure 5.3). Variation is less pronounced with shoot-tip culture but increases markedly with callus and isolated cell culture. Clearly more research on variants produced in shoot-tip and bud cultures is warranted before field gene banks are reduced in scale or scope in favor of short-term *in vitro* collections.

Germplasm preservation in tissue cultures will become much more effective when long-term storage techniques are perfected. Preservation at ultra-low temperatures of around − 196° C, commonly called cryopreservation, is being explored as a no-growth technique for germplasm storage. At such low temperatures, biological activity essentially ceases, so no genetic changes should occur during storage. The main problems are survival of shoot tips and buds under such low temperatures and regeneration of vigorous plants. Embryonic tissues of date palm and African oil palm have survived liquid nitrogen temperatures for eight months (Wetter, 1984). Cryoprotectants and quick freezing are used to help place delicate germplasm into ultra-low temperature storage. Most cryoprotectants act essentially as antifreeze agents for plant tissues.

Cryogenic storage will help reduce time-related genetic change. Long-term experiments are needed to determine whether genetic alterations occur under cryopreservation and the IBPGR has initiated a research project in this area. Experiments thus far have been successful for two-year storage of potato and chickpea (Bajaj, 1983). Although difficult to handle under cryogenic storage, shoot-tip cultures are preferred for long-term storage because they will be more stable, are easier

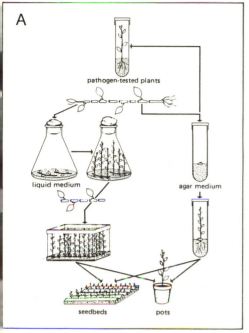
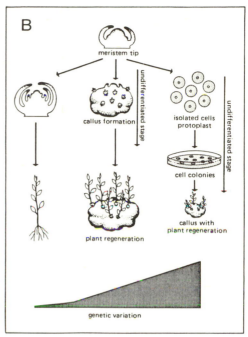

5.3. Tissue culture in potato and other root crops has become an important and routine tool: A. *In vitro* propagation method for potato. B. Genetic variation in potato under different *in vitro* procedures. (Courtesy of the International Potato Center.)

to regenerate into plants, and they can be used to produce disease-free material.

GERMPLASM EXCHANGE

Shoot-tip cultures are rapidly becoming the preferred tissue for the international exchange of clonal material (Withers, 1982). A major reason for this development is that tissue culture can produce disease-free germplasm, particularly with respect to viruses. Freeing otherwise healthy plants of viruses usually involves removal of the shoot tip at the apical meristem plus one or two leaf primordia. Rapidly growing material has less chance of being invaded by viruses than older cells.

At CIAT in Colombia, *in vitro* cultures for exchanging or storing cassava germplasm are prepared by first cutting stakes from healthy-appearing plants. Stakes are then grown out in a glasshouse to see if any latent diseases manifest themselves. Duplicate stakes are placed in a

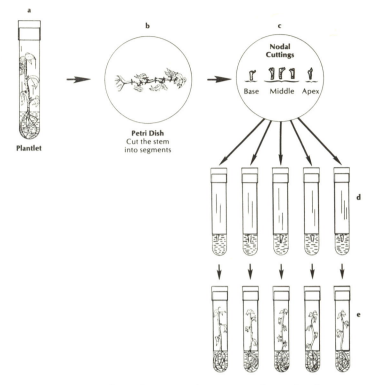

a

b

c

Nodal
Cuttings

Base Middle Apex

Petri Dish
Cut the stem
into segments

Plantlet

d

e

5.4. Procedures for preparing cassava clones for tissue-culture storage and exchange at Centro Internacional de Agricultura Tropical near Cali, Colombia.
(From Roca et al., 1984.)

heat chamber for three weeks where the diurnal temperature is a constant 40° C and the nocturnal temperature is 35° C. Shoot tips are taken
from the heat-treated cassava stakes and are cultured to produce a
plantlet. After a month's growth, the plantlet is divided into single-node
cuttings (Figure 5.4). For storage, five cuttings are cultured per accession. Follow-up serological tests are performed to check for viruses
(IBPGR, 1983b:3). From 1978 to 1984, CIAT received 1,588 cassava
accessions in tissue-culture form from other collections (CIAT,
1985:39). *In vitro* disease elimination techniques provide a powerful
new tool to ensure safe international exchange of germplasm, particularly for vegetatively propagated crops or for plants in which viral
transmission through seed is known to occur.

In the future, all nations exchanging germplasm will need special as-
ceptic facilities for sterile transfer of plant germplasm. Pathogen-free

germplasm depends not only on sound tissue culture techniques but also on trustworthy methods of disease indexing (identification and diagnosis). For some crops, disease-indexing methods are available, but for others they have yet to be developed. Monoclonal antibodies, currently one of the most useful products of recombinant DNA research, promise to be a major tool for checking the presence of pathogens in germplasm in the near future.

The international agricultural research centers are already using *in vitro* methods to send elite materials (advanced and improved lines) to colleagues around the world. As of early 1985, for example, CIAT has sent fifty elite cassava varieties to countries in Latin America and Southeast Asia (CIAT, 1985:39). Germplasm of vegetatively propagated crops is not only cheaper to maintain in tissue culture than in field plantings, mainly because of reduced labor costs, but it is also less expensive to ship because it is lighter than tubers or cuttings. Another benefit is the generally increased yield of cassava plants grown from tissue culture, probably because they are freer of diseases (CIAT, 1984a).

In spite of breakthroughs in test-tube maintenance of germplasm and indexing of diseases in tissue cultures, out-of-date quarantine procedures hamper the effectiveness of *in vitro* germplasm exchange. Many quarantine stations are not equipped nor knowledgeable enough about *in vitro* techniques to handle and process test-tube materials safely. Quick, inexpensive, and accurate disease indexing is especially needed under such circumstances; only in this manner can control regulations be satisfied and valuable germplasm be allowed to reach the hands of breeders.

In addition to germplasm exchange and storage, *in vitro* techniques will find a place in collecting vegetatively propagated material in the field as well as in preserving some species with recalcitrant seeds. Shoot-tip and bud cuttings, promptly placed in *in vitro* cultures after preliminary cleaning, are being tested on cacao and cassava with a view to using such techniques for gathering germplasm in the field. Work is in progress on coconut embryos which are easier to handle than the bulky nuts. *In vitro* collecting methods are inexpensive and easy to use, and therefore will be widely adopted.

GENETIC RESOURCE CONTRIBUTIONS TO BIOTECHNOLOGY

Is the biotechnology/genetic resource relationship just a one-way street—that of biotechnological contributions to genetic-resources work and plant breeding? Some publications on biotechnology seem to say as much, but the truth is that gene banks are a basic resource for biotechnology. Lowell N. Lewis (1985) of the University of California expressed the relationship thus: "An adequate gene resource conservation program is to genetic engineering as a library is to knowledge. They both are sources of the past, the present, and the future, and both are equally essential to the development of their counterpart." Germplasm collections of crops, their wild relatives, and unrelated plants are essential if biotechnology is to move ahead (Witt, 1985:62).

Some of the publicity surrounding genetic engineering depicts recombinant DNA researchers creating new genes, thus rendering germplasm collections obsolete. This is unlikely for two reasons. First, biotechnologists need a model in order to synthesize a gene. At the moment at least, biotechnologists cannot invent genes; genes can only be shifted between certain organisms and copied. Second, genetic engineers will continue to depend heavily on naturally occurring genes in their experiments for the foreseeable future. Gene banks are thus destined to provide important recruiting grounds for gene hunters.

Biotechnology relies heavily on gene-bank curators to tell them what germplasm is available, its characteristics and problems, where collections are held, and how to gain access to them. For many crops, the main contribution of biotechnology for the next ten years or so will be in the use of major crop plants as research subjects. Such research will ensure new information on major crops and will probably help to overcome problems such as the lack of effective DNA vector systems for monocots and reliable protoplast-fusion techniques for major cereals, root crops, and grain legumes.

For some biotechnology work, the availability of a reliable and well-prepared catalog of germplasm holdings is likely to be the most important crop-related reference. Gene banks must stand ready to backstop biotechnologists with evaluated germplasm that presents an array of characters useful in molecular and cellular studies (Table 5.1). Crop scientists can assist biotechnologists by providing information on major production problems and by cooperating in whole plant evaluation.

The future growth of biotechnology research rests upon a team approach that enlists various disciplines. Just as biotechnologists bring special skills in molecular and cell biology as well as in biochemistry, crop scientists contribute information about the whole plant and the competitive environment of crops in the field. Agricultural scientists also supply crucial knowledge about suitable agronomic conditions for the crop, including soil and water factors, as well as data about weeds and the biology and population dynamics of pests and diseases. Crop scientists also provide important inputs with respect to plant genetics,

TABLE 5.1

Indigenous rice varieties with special traits held in gene banks in national programs and at the International Rice Research Institute.

Traits	Samples	Traits	Samples
Salinity tolerant	345	Cool tolerant	671
Aluminum tolerant	290	Nematode resistant	5
Acidity tolerant	216	Disease resistant	1,634
Alkaline tolerant	324	Insect resistant	633
Iron tolerant	16	Aromatic types	122
Iron-deficiency tolerant	3	Semi-dwarfs	84
Phosphorus-deficiency		Rodent resistant	10
tolerant	3	Medicinal uses	13
Drought resistant	2,807		
Flood tolerant	776	Total	8,025

Source: IRRI, 1980.

selection, and germplasm characteristics, as well as field research methods and statistical analyses that are needed to assess crop plants under field conditions. Genetic engineers will need the assistance of competent plant breeders to obtain finished products that are agronomically suitable and economically advantageous (Chang, 1984b).

GENES IN THE BANK

In this chapter we present a general status report on germplasm collections for the important crops. Our approach is necessarily statistical, but we also analyze collection and evaluation gaps and the locations of gene banks. We wish to emphasize that large numbers of accessions do not necessarily indicate that a crop has been adequately collected. The degree of overlap in collections and of comprehensiveness varies considerably among crops. Furthermore, the size of collections and their degree of coverage of landraces and wild relatives are only rough estimates made by staff of the International Board for Plant Genetic Resources (IBPGR), gene-bank curators, and plant breeders. Nevertheless, as crude as the available figures are, they are useful for gaining an idea of how far germplasm collecting of different crops has proceeded.

Over 2.5 million crop accessions are held in germplasm collections throughout the world, including over 1.2 million accessions of cereals, 369,000 accessions of food legumes, 215,000 accessions of forage legumes and grasses, 137,000 accessions of vegetables, and 74,000 clones of root crops (Table 6.1). Crops of major economic importance that are backed by strong agricultural research programs are best represented in gene banks (Lyman, 1984). Cereals account for the bulk of international food shipments in both volume and value and therefore, not surprisingly, make up the largest category of samples held in gene banks. Grain legumes (pulses) are also commercially important, and large numbers are kept in numerous gene banks at diverse locations. And the large number of forage legume and grass accessions is a reflection of the importance of livestock both locally and on international markets.

The sizable overlap in accessions for each crop is evident from the distinct-samples column of Table 6.1. For most crops, at least half of the combined germplasm collections are replicate accessions (Lyman, 1984). More than two-thirds of the world's wheat germplasm accessions are replicated. It is IBPGR policy for the designated base collections (long-term storage for security) to hold duplicates of each other's material as insurance against loss from natural disasters, human error, or political upheaval. But indiscriminate duplication of a given species

TABLE 6.1
Estimated number of gene-bank accessions worldwide.

	Number of			Coverage Percentages	
Crop	Accessions	Distinct Samples	Collections of 200+ Accessions	Landraces	Wild Spp
CEREALS					
Wheat	410,000	125,000	37	95	60
Barley	280,000	55,000	51	85	20
Rice	215,000	90,000	29	75	10
Maize	100,000	50,000	34	95	15
Sorghum	95,000	30,000	28	80	10
Oats	37,000	15,000	22	90	50
Pearl millet	31,500	15,500	10	80	10
Finger millet	9,000	3,000	8	60	10
Other millets	16,500	5,000	8	45	2
Rye	18,000	8,000	17	80	30
PULSES					
Phaseolus	105,500	40,000	22	50	10
Soybean	100,000	18,000	28	60	30
Groundnut	34,000	11,000	7	70	50
Chickpea	25,000	13,500	15	80	10
Pigeonpea	22,000	11,000	10	85	10
Pea	20,500	6,500	11	70	10
Cowpea	20,000	12,000	12	75	1
Mungbean	16,000	7,500	10	60	5
Lentil	13,500	5,500	11	70	10
Faba bean	10,000	5,000	10	75	15
Lupin	3,500	2,000	8	50	5
ROOT CROPS					
Potato	42,000	30,000	28	95	40
Cassava	14,000	6,000	14	35	5
Yams	10,000	5,000	12	40	5
Sweet potato	8,000	5,000	27	50	1
VEGETABLES					
Tomato	32,000	10,000	28	90	70
Cucurbits	30,000	15,000	23	50	30
Cruciferae	30,000	15,000	32	60	25
Capsicum	23,000	10,000	20	80	40
Allium	10,500	5,000	14	70	20
Amaranths	5,000	3,000	8	95	10
Okra	3,600	2,000	4	60	10
Eggplant	3,500	2,000	10	50	30

TABLE 6.1 (*cont.*)

Crop	Number of			Coverage Percentages	
	Accessions	Distinct Samples	Collections of 200 + Accessions	Landraces	Wild Spp
INDUSTRIAL CROPS					
Cotton	30,000	8,000	12	75	20
Sugar cane	23,000	8,000	12	70	5
Cacao	5,000	1,500	12	*	*
Beet	5,000	3,000	8	50	10
FORAGES					
Legumes	130,000	n.a.	47	n.a.	n.a.
Grasses	85,000	n.a.	44	n.a.	n.a.

Notes: The coverage percentages are estimates derived from a consensus of scientific opinion (Lyman, 1984). No precise figures can be given until all accessions have been described. The coverage of wild species relates to those largely in the primary gene pool—that is, those species that were progenitors of crops, have co-evolved with cultivated species by continuously exchanging genes, or are otherwise closely related. Such wild species are more easily used in breeding. In addition, hundreds of more distantly related species, with great potential for crop improvement, exist in secondary or tertiary gene pools, but they have not been adequately considered. Estimates of these are not provided.
* Coverage difficult to estimate because many selections are from the wild.
n.a. = data not available.

within the same collection, or maintenance of the entire collection of a species at several places, is costly and unnecessary. Gene-bank operators, particularly at the international agricultural research centers, have a major task in eliminating constantly redundant samples to streamline collections and make them more manageable.

With respect to the status of collections, we will discuss the size, storage conditions, and location of germplasm assemblages of cereal, root, food legume, and industrial crops. We chose an arbitrary cut-off of one thousand accessions in order to keep tables manageable. Figures for gene-bank accessions are only approximate since material is constantly being added or transferred. Also, some of the facilities listed with medium- or long-term facilities may operate intermittently or not at all because of financial or technical constraints. Furthermore, because of inadequate drying, cooling, and regeneration at certain locations, the viability of many accessions is questionable. This partly stems from the fact that many collections now considered conservation depositories arose from the efforts of scientists whose goals were not conservation.

Nevertheless, accession lists provide a useful framework for analyzing the location of holdings and the evaluation and collection gaps.

CEREALS

Cereals in general are the best-collected crops, and wheat tops the list in numbers of accessions and the degree of comprehensiveness. Wheat, domesticated in the Middle East, accounts for some 410,000 accessions in close to forty gene banks. Most wheat landraces have been collected, as have 60 percent of the germplasm of wild relatives. Except for those in a few scattered pockets, wheat landraces were essentially all collected by the end of 1985 as anticipated (IBPGR, 1984b). The major collections for long-term storage of wheat germplasm are concentrated in the industrial nations, particularly at the Vavilov All-Union Institute of Plant Industry (VIR) in Leningrad, the National Seed Storage Laboratory (NSSL) at Fort Collins, Colorado, and the Istituto del Germoplasma in Bari, Italy (Table 6.2). The Plant Germplasm Institute at the University of Kyoto, Japan, maintains a base collection of wheat relatives, mainly species of *Triticum* and *Aegilops*. Materials from these collections are available, and IBPGR recently helped repatriate material to developing countries that had lost their own collections.

In the Third World, the largest wheat gene banks are found at the International Maize and Wheat Improvment Center (CIMMYT—Centro Internacional de Mejoramiento de Maiz y Trigo) near Mexico City, and at the International Center for Agricultural Research in the Dry Areas (ICARDA) near Aleppo, Syria. CIMMYT maintains a medium-term collection of 31,000 tropical and subtropical wheats and triticale (a cross between wheat and rye), while ICARDA holds 16,500 wheat samples. With the help of the Japanese government, CIMMYT completed construction of facilities for medium-term storage of wheat in 1982, thereby increasing the number of such facilities for the crop in the developing world to five (Table 6.2). ICARDA is building a medium-term storage facility for its wheat germplasm collection. Wheat breeders throughout the world still rely heavily on the small-grains collection at Beltsville, Maryland, for material.

Barley (*Hordeum vulgare*), with some 280,000 accessions worldwide, is the second best-collected crop in number of gene-bank samples (Table 6.1). This ancient cereal, which featured in the New Testament story about the feeding of the five thousand and in some ancient cultures symbolizes immortality or rebirth, was domesticated in the Middle East and is relatively well represented in gene banks. Most accessions are held in temperate countries (Table 6.3), where the crop is used mostly by brewers and to feed livestock; stubble left after harvesting the grain is traditionally grazed by unpenned flocks of sheep and goats in the Mediterranean region. In the tropical highlands and in the drier parts of the subtropics, people also eat barley in soups and porridges. The Centro Nacional de Pesquisa de Trigo (CNPT) in Brazil, ICARDA in Syria, and CIMMYT in Mexico hold the largest barley germplasm collections in the Third World. Some barleys tolerate drought, frost, and saline soils and can be used in wide crosses with other cereals, particularly rye and wheat. Some landraces remain to be collected in tropical highlands and only one-fifth of the germplasm of barley relatives is in gene banks (Table 6.1), due in part to the large number of wild barley species and their widespread distribution.

TABLE 6.2
Wheat (*Triticum* spp.) in gene banks.

No. of Accessions	Type of Storage	Institution	Location
74,500	M, L	VIR*	Leningrad, USSR
39,003	M	USDA	Beltsville, Md., USA
37,477	L	NSSL*	Fort Collins, Colo., USA
31,144	M	CIMMYT	El Batán, Mexico
31,000	S	ARO	Bet Dagan, Israel
26,000	M, L	IG*	Bari, Italy
22,100	L	NSWDA	Tamworth, Australia
20,000	S#	CGI	Beijing, China
16,596	S#	ICARDA	Aleppo, Syria
16,000	S	IARI	New Delhi, India
13,600	S#	IPIGR	Plovdiv, Bulgaria
10,875	M, L	FAL	Braunschweig, F.R. Germany
10,000	M, L	ZGK	Gatersleben, German D.R.
8,000	S#	IHAR	Radzikow, Poland
7,201	M, L	PGRC	Addis Ababa, Ethiopia
7,000	S	CNPT	Passo Fundo, Brazil
6,774	M	PGI	Kyoto, Japan
6,000	S	RICTP	Fundulea, Romania
6,000	M	PARC	Islamabad, Pakistan
5,000	M, L	IGPB	Prague, Czechoslovakia
4,852	M, L	INTA	Pergamino, Argentina
4,506	M, L	PBI	Cambridge, UK
4,200	M, L	NIAS	Tsukuba, Japan
4,000	S	SPA	Wukung, China
4,000	M	ARARI	Menemen, Turkey
4,000	M	SVP	Wageningen, Netherlands
4,000	M	NIAVT	Tapioszele, Hungary

TABLE 6.2 (*cont.*)

No. of Accessions	Type of Storage	Institution	Location
2,500	S	INRA	Versailles, France
2,000	S	UC	Riverside, Calif., USA
1,726	M	DARS	Kabul, Afghanistan
1,221	S	NBPGR	New Delhi, India
1,200	M	UNA	Lima, Peru

Sources: The authors' field notes, plus Bhatti *et al.*, 1983; Chang, 1985; Chen, 1983; Chitrakon *et al.*, 1983; *Crop Germplasm Conservation and Use in China* (Rockefeller Foundation, New York, 1980); Esquinas-Alcazar, 1982; *Plant Genetic Resources Newsletter* 49:13 (1982); Faris, 1984; Hawkes, 1985; IBPGR, 1980, 1985d; IBPGR, Regional Committee for Southeast Asia, *Newsletter* 4(2):7 (1980), and 4(3):5 (1980); ICARDA, 1984:15: *IITA Research Briefs* (International Institute of Tropical Agriculture, Ibadan, Nigeria), Vol. 6 (2):4-5 (1985); Johnson and Beemer, 1977; Kyaw, 1983; McDonald, 1984; Mengesha, 1984; Ng et al., 1983; Tay et al., 1984; Valls, 1985; Xuan and Luat, 1983; personal communications from L. N. Bass, D. Bondioli, F. Cardenas Ramos, L. Holly, F. E. Lopez, G. R. Lovell, K. C. Nagel, W. H. Skrdla, D. H. Smith, and J. Wynne.

Notes: Storage-temperature ranges are: short-term (S), 6°C to ambient; medium-term (M), 0° to 5°C; long-term (L), −10° to −20°C. Most of the short- and medium-term collections have been, or will be, put into long-term storage at appropriate institutions.

See Appendix 2 for acronyms and Chapter 4 for additional information on germplasm-storage conditions.

* IBPGR-designated base collection.

Long-term facility under construction.

Close behind barley in number of accessions, but of greater importance in human nutrition and commerce, is rice. With a global total of about 215,000 accessions, rice is the third best-represented crop in gene banks. It is a basic staple in much of the Third World, particularly in Asia, where common rice (*Oryza sativa*) was domesticated, and seven of the ten largest rice gene banks are in developing countries (Table 6.4). The main gene banks for tropical rices are located at the International Rice Research Institute (IRRI) in the Philippines, the Central Rice Research Institute in India, the Central Research Institute for Food Crops in Bogor, Indonesia, the International Institute of Tropical Agriculture (IITA) in Nigeria, and the West Africa Rice Development Association (WARDA) in Liberia. Japan and the United States maintain major collections of temperate rices and act as back-up security for the IRRI and IITA materials.

Since its inception in 1960, IRRI has assembled the world's largest rice collection, containing more than twice as many accessions as the second largest germplasm collection for the cereal. Of the 78,800 accessions in IRRI's gene bank in mid-1985, 73,300 are *Oryza sativa*, 2,900 are African rice (*O. glaberrima*), 1,900 are wild species, and 700 are genetic testers and mutants (T. T. Chang, pers. comm.). The International Rice Germplasm Center at IRRI is the largest germplasm collection for any crop and is regarded as one of the best-managed gene banks and a model for future efforts. Much of the credit for organizing and carefully maintaining IRRI's germplasm collection belongs to T. T. Chang (Figure 4.2), a geneticist born in mainland China and trained in the United States. In recognition for his dedication to germplasm con-

TABLE 6.3
Barley (*Hordeum* spp.) in gene banks.

No. of Accessions	Type of Storage	Institution	Location
25,284	L	NSSL	Fort Collins, Colo., USA
23,371	M	USDA	Beltsville, Md., USA
21,000	L	PGRO*	Ottawa, Canada
19,500	S	CNPT	Passo Fundo, Brazil
17,459	M, L	VIR	Leningrad, USSR
14,215	M#	ICARDA	Aleppo, Syria
13,900	L	NGB*	Lund, Sweden
10,200	M, L	ZGK	Gatersleben, German D.R.
6,025	M	BGC	Kurashiki, Japan
5,569	M	CIMMYT	El Batán, Mexico
5,263	M, L	NIAS*	Tsukuba, Japan
5,017	M, L	FAL	Braunschweig, F. R. Germany
5,000	M, L	PGRC*	Addis Ababa, Ethiopia
4,900	M, L	PBI	Cambridge, UK
4,500	S#	IHAR	Radzihow, Poland
3,500	S	RICTP	Fundulea, Romania
3,200	S#	IPIGR	Plovdiv, Bulgaria
3,100	S	SCRI	Edinburgh, UK
3,000	S	IARI	New Delhi, India
2,600	M	PARC	Islamabad, Pakistan
2,500	S	JNRC	Lincoln, UK
2,500	M, L	RIPP	Praha, Czechoslovakia
2,300	M	SVP	Wageningen, Netherlands
2,000	M	CSIRO	Canberra, Australia
1,500	M	UNA	Lima, Peru
1,460	M	RIPP	Bratislavska, Czechoslovakia
1,333	M	GGB	Thessaloniki, Greece
1,275	S	CNIA	Buenos Aires, Argentina

TABLE 6.3 (*cont.*)

No. of Accessions	Type of Storage	Institution	Location
1,240	S, M	IPB	Jokioinen, Finland
1,200	S	IARI	New Delhi, India
1,000	S	ARO	Bet Dagan, Israel
1,000	M, L	IG	Bari, Italy

Note: See notes to Table 6.2.
* IBPGR-designated base collection. # Long-term facility under construction.

TABLE 6.4
Rice (*Oryza* supp.) in gene banks.

No. of Accessions	Type of Storage	Institution	Location
78,800	M, L	IRRI*	Los Baños, Philippines
18,065	L	NSSL*	Fort Collins, Colo., USA
18,000	M, L	NIAS*	Tsukuba, Japan
13,050	S	CRRI	Cuttack, India
13,511	S	CRIFC	Bogor, Indonesia
11,230	S, M	USDA	Beltsville, Md., USA
8,600	M, L	IITA*	Ibadan, Nigeria
8,226	M	WARDA	Monrovia, Liberia
6,675	M, L	CENARGEN	Brasília, Brazil
6,000	S, M. L	RRI	Bangkok, Thailand
5,100	M	BRRI	Dacca, Bangladesh
3,842	M	IRAT	Montpellier, France
3,500	M, L	VIR	Krasnodar, USSR
3,200	S	INTA	Córdoba, Argentina
3,130	M, L	KSB	Penang, Malaysia
2,745	S	CARI	Peradeniya, Sri Lanka
2,500	S#	CGI	Beijing, China
2,080	S	ARI	Yezen, Burma
2,000	S	IRA	Tananarive, Madagascar
1,700	S, M	TARI	Taipei, Taiwan, China
1,500	S	NIAS	Hanoi, Vietnam
1,419	M	PARC	Islamabad, Pakistan
1,400	S	CARS	Lilongwe, Malawi
1,027	M, L	AES	Suweon, Rep. of Korea
1,000	S	ICA	Bogotá, Colombia
1,000	S	ORSTOM	Paris, France

Note: See notes to Table 6.2.
* IBPGR-designated base collection.   # Long-term facility under construction.

servation, Dr. Chang has been honored by the American Society of Agronomy and the Institute of Biology, London, and was recently appointed principal scientist at IRRI.

Computerized information on IRRI's rice collection is especially useful because of the meticulous attention paid to the recording of data on samples. Forty-five morphological and agronomic characteristics are recorded for each entry. Entomologists, plant pathologists, physiologists, chemists, and soil scientists at IRRI add as many as thirty-eight genetic evaluation and utilization (GEU) traits, covering resistances to diseases and pests as well as tolerances to adverse soils and climates to the data files (Chang, 1984c). Because of IRRI's commitment to operating an efficient gene bank with evaluated accessions, breeders of rice probably know more about the genetic resources of their crop than other crop breeders.

As of mid-1985, 22,000 samples were in IRRI's medium-term storage facility and an equal number in the long-term cold room. The germplasm collection is gradually regenerated, and fresh seed is put in medium- and long-term storage. Older seed stocks in medium-term storage (2° C) are in sealed glass bottles containing silica gel, but new seed stocks are being vacuum-packed in aluminum cans. A duplicate set of the canned accessions is deposited with NSSL in Fort Collins for added security.

Since 1980 IRRI has implemented a systematic approach toward long-term seed preservation. For long-term storage at $-10°$ C, the moisture content of seeds is reduced to 6 percent before the accession is preserved in two aluminum cans, each containing 2,500 to 4,000 seeds. About 20,000 rice varieties and much wild material remain to be collected. Major areas with landraces still to be collected are the border zones of Kampuchea, Laos, Vietnam, and Thailand. Collecting has been slow in many of these border areas because of difficult access and political turmoil. In Thailand, for example, thousands of traditional varieties await collection in the north, parts of the southwest, and along the border with Burma and Laos (Chitrakon et al., 1983). Because of the great genetic richness of cultivated and wild rices and segments of Chinese and Indian germplasm collections yet to be deposited with IRRI, the Institute's gene bank is expected to continue growing until it reaches about 130,000 accessions (Chang, 1984a; T. T. Chang, pers. comm.).

The rice gene bank at IITA in Ibadan, Nigeria, holds 6,500 accessions of *Oryza sativa*, mostly from Africa, in medium- and long-term storage. With IBPGR support, IITA has also assembled 2,000 samples of the other cultivated rice, *Oryza glaberrima*, which is indigenous to Africa. IRRI also has accessions of this rice, which is losing ground to the Asian domesticate *O. sativa*. The two species occasionally hybridize when grown close together and thereby exchange genes. IITA assists WARDA in its germplasm work, which concentrates on acquiring and improving African paddy rices, now totaling 8,000 entries. Upland rices will receive increased attention at IITA and WARDA.

Maize, with about 100,000 gene-bank accessions worldwide and 95 percent of landraces collected, does not appear to be such a genetically diverse crop as rice (Table 6.1). The fact that maize is an outcrossing crop, rather than a self-pollinator such as rice or wheat, probably accounts for much of the reduced number of landraces. Maize is a valuable commercial crop in North America, where it is used principally as livestock feed, for syrup (used to sweeten many soft drinks), to make breakfast cereals and snack foods, and to produce ethanol for mixing with gasoline, known popularly as gasohol. In the Third World, and especially in Latin America and Africa, maize is a basic staple; it is eaten fresh or as a steamed mash (polenta), ground into flour to make tortillas or chapatis, and fed to pigs and chickens. Domesticated in Mexico, it shows secondary diversity in the Andes and to a lesser extent in southern Europe and the Mediterranean region. In parts of Asia such as India and the Philippines, the crop has been eaten principally as green corn for centuries, and its cultivation is spreading.

The dispersed location of gene banks for maize reflects its global importance (Table 6.5). The largest maize germplasm collections are in the Soviet Union and Yugoslavia, where 30,000 accessions are housed under medium- and long-term storage. In the United States, the University of Illinois serves as a repository of maize mutants and holds more than 100,000 samples for use by the world research community (G. B. Fletcher, pers. comm.).

Long-term gene banks for maize are located both in industrial and developing nations. NSSL in Fort Collins contains a base collection of New World maize; the Vavilov Institute in Leningrad concentrates on European varieties; the Maize Institute at Braga, Portugal, has about one thousand cultivars of the cereal, mostly from Iberia and other parts of southern Europe; and the National Institute of Agricultural Sciences at Tsukuba, Japan, maintains a long-term collection of Asian maize. In the Third World, CIMMYT's long-term gene bank in Mexico for maize became operational in early 1985, and Argentina and the Philippines maintain 3,000 and 1,678 accessions, respectively, under long-term conditions. The Thailand Institute of Scientific and Technological Research (TISTR) recently inaugurated a long-term maize gene bank in Bangkok. TISTR is acquiring Asian maize and forms part of IBPGR's

119

TABLE 6.5

Maize (*Zea mays*) in gene banks.

No. of Accessions	Type of Storage	Institution	Location
15,084	M, L	VIR*	Leningrad,USSR
15,000	M	IMR	Belgrade, Yugoslavia
11,100	M, L	CIMMYT	El Batán, Mexico
10,000	M[#]	INIA	Chapingo, Mexico
7,619	L	NSSL*	Fort Collins, Colo., USA
7,145	M	UNA	Lima, Peru
5,000	S	ICA	Medellin, Colombia
3,200	S	RICTP	Fundulea, Romania
3,000	M, L	INTA	Pergamino, Argentina
3,000	M	ISU	Ames, Iowa, USA
2,800	L	PGRO	Ottawa, Canada
2,654	M, L	NIAS*	Tsukuba, Japan
2,220	M	CIFEP	Cochamamba, Bolivia
1,678	M, L	IPB	Los Baños, Philippines
1,571	S[#]	IARI	New Delhi, India
1,500	M	NARS	Kitale, Kenya
1,368	M	CRIFC	Sukamandi, Indonesia
1,306	M	MRI	Trnava, Czechoslovakia
1,040	M, L	INIA	Madrid, Spain
1,000	S	CNU	Daejeon, Rep. of Korea
1,000	M[#]	MI	Braga, Portugal

Note: See notes to Table 6.2.

* IBPGR-designated base collection.   [#] Long-term facility under construction.

network of base collections. The Mexican national program, INIA (Instituto Nacional de Investigaciones Agricolas), is building a long-term storage facility for germplasm of maize and other seed crops at Zacatecas. At an elevation of 2,200 meters and with only 300 millimeters of annual rainfall, the energy costs of maintaining germplasm samples at Zacatecas should be reasonable.

CIMMYT and INIA maintain their medium-term collections of maize about an hour's drive northeast of Mexico City, in El Batán and Chapingo (Table 6.5). CIMMYT's maize collection contains samples from over fifty Latin American and Caribbean countries. CIMMYT regenerates some five hundred accessions a year, and Pioneer Hi-Bred is collaborating with CIMMYT and the U.S. Department of Agriculture to regenerate some of them. Pioneer Hi-Bred, in turn, duplicates material of interest to its research programs. Landraces comprise most of

TABLE 6.6
Sorghum (*Sorghum bicolor*) in gene banks.

No. of Accessions	Type of Storage	Institution	Location
24,600	M, L	ICRISAT*	Hyderabad, India
14,000	L	NSSL*	Fort Collins, USA
9,815	S	SRPIS	Experiment, Ga., USA
9,615	M, L	VIR	Leningrad, USSR
5,000	M, L	PGRC	Addis Ababa, Ethiopia
4,900	S	RICTP	Fundulea, Romania
4,610	S	USSCFS	Meridian, Mississippi, USA
4,000	S	MITA	Mayaguez, Puerto Rico
4,000	S	ASP	Tihama, Yemen
3,000	S#	CGI	Beijing,China
3,000	M	INIA	Chapingo, Mexico
2,700	S	INTA	Córdoba, Argentina
2,626	M	ORSTOM	Bondy, France
2,072	M, L	IPB	Los Baños, Philippines
2,000	S#	IARI	New Delhi, India
1,500	S	KU	Bangkok, Thailand
1,000	S	CSIRO	St. Lucia, Australia

Note: See notes to Table 6.2.
* IBPGR-designated base collection. # Long-term facility under construction.

CIMMYT's maize collection, but wild relatives such as annual teosinte (*Zea mexicana*) and the recently discovered perennial corn (*Z. diploperennis*) are also included.

The collection of maize landraces in Latin America and the Mediterranean was essentially completed by 1986. Most of the estimated 5 percent of landraces remaining to be collected are in Asia and Africa (IBPGR, 1984b). Although a good proportion of the traditional varieties of maize are in gene banks, many of the older collections may no longer be viable (Lyman, 1984). Furthermore, only a tiny fraction of the germplasm of maize relatives is safely in gene banks.

Sorghum is used mainly for livestock feed and to manufacture syrup in the industrial nations, but in the drier regions of Africa and India it is a food for millions of people. Sorghum is increasingly imported by developing countries, such as Mexico, for livestock feed, particularly for poultry. Despite its widespread and growing importance as a food, sorghum is unevenly represented in gene banks (Table 6.6). The global total for sorghum germplasm accessions is around 95,000, with the International Crops Research Institute for the Semi-Arid Tropics

6.1. The Plant Genetic Resources Center in Addis
Ababa, Ethiopia, has long-term storage facilities for wheat,
barley, sorghum, and minor millet germplasm. 1984.

(ICRISAT) near Hyderabad, India, holding close to a quarter of that
total. ICRISAT's 24,000 sorghum samples come from sixty-eight countries and are kept under medium- and long-term conditions; the collection has grown markedly since 1978, when it had only 15,000 entries
(ICRISAT, 1978). Of the four long-term collections of sorghum, three
are in developing countries: India, Ethiopia (Figure 6.1), and the Philippines (Table 6.6). ICRISAT inaugurated its long-term storage facilities with the help of the Japanese government and the Asian Development Bank in 1983.

The millets, comprising some dozen species in six genera, seldom enter world trade but are nevertheless a valuable human food in arid por-
tions of Africa, Asia, and the Mediterranean. Millets thrive in ecologically marginal lands, such as areas with poor soils or regions subject to
drought, hence their importance to millions of people in developing
countries. Only a handful of gene banks contain sizable millet germplasm collections, and accessions for all species do not exceed 57,000
(Table 6.1). Pearl millet is the best-represented millet in gene banks;
ICRISAT has a collection of 14,000 accessions gathered in twenty-five
countries (Table 6.7). ICRISAT also has the most diverse collection of
millet species under long-term storage.

IBPGR has emphasized the collecting of landraces of pearl millet,
and impressive strides have been made in gathering traditional varieties

in Africa during the last decade. Only about half of the varieties of the minor millets are in gene banks, and collecting for those crops is expected to continue well into the 1990s. Teff (*Eragrostis tef*), for example, is a cultivated cereal native to Ethiopia and is used to make a thick, savory pancake-like bread called *ingera*. Dark brown-colored *ingera* is a basic staple; it is eaten alone or lines the plate, to which lamb, cheese, and other dishes are added. Teff is exceptionally high in protein, approaching 15 percent, but it has very small grains. The Plant Genetic Resources Center in Addis Ababa holds only a small germplasm collection, and more extensive collections are needed to provide materials for improving the yield of this highly nutritious crop.

The collection of landraces of wheat, maize, barley, and oats is expected to be nearly complete by 1986, rice by 1988, and sorghum and pearl millet by the late 1980s. Some durum wheat landraces in stressful environments and primitive bread wheats still need to be collected in parts of North Africa, southern Europe and the Soviet Union, as well as in Oman, Saudi Arabia, Iraq, Yemen, and Lebanon (Hawkes, 1985:12). Collecting work is difficult and often impossible, in some cases due to armed conflict. Because expanding areas are planted with high-yielding cereal varieties, landraces are becoming scarce in many countries, even in remote areas; the number of samples collected per mission is decreasing, as it is with the collecting of wild relatives.

In the Andes, some of the minor grain crops are losing ground to the major cereals, especially to modern cultivars of wheat and barley. Two chenopod species, quinoa (*Chenopodium quinoa*) (Figure 6.2) and cañahua (*C. pallidicaule*), for example, are still sown from highland Ecuador south to Chile, but the cultivated area devoted to these cereal-like crops has shrunk since the Spaniards introduced Old World cereals in the early 1500s. If adequate germplasm collections are built up, breeders may eventually be able to upgrade the yield of quinoa and other minor crops so that they can better face the challenge of modern cultivars of the major crops.

Several Andean countries have started to conserve the germplasm of chenopods and other cultigens of local and regional importance. The Universidad Nacional Técnica del Altiplano, in Puno, Peru, for example, has the largest collection of chenopod germplasm, with 2,200 accessions of *C. quinoa* and 420 of *C. pallidicaule*. Also at Puno, the national agricultural research program maintains a working collection of 1,820 *C. quinoa* and 222 *C. pallidicaule* accessions. And in Lima, the Universidad Nacional Agraria La Molina has a medium-term collection of 1,500 entries of quinoa.

The decline of chenopods is lamentable, for these crops help stamp

TABLE 6.7
Millet in gene banks.

Species/ Accessions	Type of Storage	Institution	Location
PEARL MILLET (*Pennisetum typhoides*)			
16,985	M, L	ICRISAT*	Hyderabad, India
2,247	S	AICMIP	Poona, India
2,100	M	ORSTOM	Bondy, France
1,200	L	PGRO*	Ottawa, Canada
1,000	L	NSSL*	Fort Collins, Colo., USA
FOXTAIL MILLET (*Setaria italica*)			
3,588	S	AICMIP	Poona, India
3,226	S	CGI	Beijing, China
1,429	S	AICMIP	Bangalore, India
1,260	M, L	ICRISAT*	Hyderabad, India
FINGER MILLET (*Eleusine coracana*)			
2,960	S	AICMIP	Bangalore, India
2,944	S	AICMIP	Poona, India
1,863	M, L	ICRISAT*	Hyderabad, India
KODO MILLET (*Paspalum scrobiculatum*)			
1,405	S	AICMIP	Poona, India

Note: See notes to Table 6.2.
* IBPGR-designated base collection.

the region's character. It is one thing, though, to espouse ecological diversity of farm lands and a wider array of dishes by promoting a rich mix of cultivated species, and another to provide farmers with economically viable options for planting. Minor crops, such as the chenopods, can only regain some of their former importance if breeding programs improve their yield, disease and pest resistance, and other characteristics. To accomplish that task, breeders will need to draw on the genetic resources of their target crops. The larger the gene pool that has been conserved, the greater the chances for success. Peru's example in assembling the genetic diversity of chenopods is therefore far-sighted. It seems unlikely that chenopods will usurp the introduced Old World cereals, but they might make a modest comeback.

The area planted to grain amaranths (*Amaranthus hypochondriacus, A. cruentus*) in Mexico has plummeted since the arrival of the Spaniards in the early 1500s. Amaranths had religious significance to the Aztecs and other groups, compelling the conquering Spaniards to suppress their cultivation vigorously because the annual crops symbolized evil and idolatry. Amaranths are now mostly a snack food in Mexico; the seeds are held together by molasses in thick, round patties called *alegria* (happiness). The Andean grain amaranth, *Amaranthus caudatus*, is also a declining relic (Sauer, 1967). Grain amaranths have lost ground to Old World cereals, but contain more protein than cereals and warrant conservation and breeding (NRC, 1984). Rodale Research Center in Kutztown, Pennsylvania, is one of the few organizations assembling amaranth germplasm. The private company has established an amaranth breeding program with a view to developing high-yielding varieties for commercial farmers.

FOOD LEGUMES

Food legumes are potentially the most valuable and yet probably the least developed of the naturally occurring sources of vegetable protein, calories, vitamins, and minerals. Some legumes, such as soybean and peanut or groundnut, are also rich in oil. Pulses have been grown for thousands of years under marginal conditions of moisture stress and low fertility in the tropics and subtropics. Unfortunately, the yield levels of most food legumes remain low and they are thus being replaced by high-yielding cereals; this trend is especially evident in India, where green-revolution wheats and rices occupy some of the land formerly planted to pulses. One manifestation of this development is India's deficit in vegetable oils, which forces the government to use valuable foreign exchange to import cooking oils.

Scientists of the Rockefeller and Ford Foundations have played leading roles in collecting the germplasm of several pulse crops in developing countries, including pigeonpea (*Cajanus cajan*) and chickpea in India, cowpea (*Vigna unguiculata*) in parts of Africa, and common bean in Latin America. With the establishment of the international agricultural research centers such as ICRISAT, IITA, ICARDA, the Asian Vegetable Research and Development Center (AVRDC), and CIAT (Centro Internacional de Agricultura Tropical) in the 1960s and 1970s, collection of the genetic variability of pulse crops accelerated. As a result, international agricultural research centers generally house large collections of pulse germplasm and most legume germplasm collections are maintained in developing countries (Table 6.8).

The degree of comprehensiveness of pulse germplasm collections ranges from 50 to 85 percent for landraces and 1 to 50 percent for wild species (Table 6.1). Two-thirds or more of the landraces of groundnut, chickpea, pigeonpea, faba bean, cowpea, lentil (*Lens culinaris*), and the common pea are in germplasm collections. The IBPGR has drawn up

6.2. Ripe quinoa (*Chenopodium quinoa*) in the Mantaro Valley, Peru, June 1982. The gentleman in the picture is Dr. Zosimo Huamán, who is in charge of the germplasm collection at the International Potato Center, Lima, Peru.

priority areas for collecting legume and other crop germplasm. Burma, Indonesia, Central America, the Caribbean, mainland China, and West Africa top the list of areas to be canvassed for groundnut germplasm, for example. In the case of soybeans, southern and western China, North Korea, and Indonesia are top-priority areas for collecting missions.

Many of the pulse crops in gene banks are the mandate crops of the international agricultural research centers, so breeders are constantly drawing on the resources of the germplasm collections. User demand helps spur collecting, and the international centers working closely with IBPGR are making steady progress in the collection and conservation of the genetic diversity of pulses. Many national programs are also collaborating in this widespread effort. Most of the landrace populations of food legumes are expected to be sampled by 1990. Only 50 to 60 percent of the landraces of soybean and *Phaseolus* beans are in gene banks; therefore the gathering of traditional varieties of those crops will probably continue into the early 1990s. CIAT expects eventually to double the size of its 35,000-sample base collection of *Phaseolus* beans (CIAT, 1985:37). As is typical for most crops, much remains to be done in preserving the germplasm of wild pulses. The collection of closely related species of food legumes is likely to extend into the next century.

ROOT CROPS

With the exception of the potato, root and tuber crops have not been as well collected as the cereals and food legumes. The relatively recent beginning of root-crop germplasm conservation and the difficulties in storing the genetic diversity of such crops are major reasons why germplasm collecting has lagged for tuber crops.

Potato, with about 42,000 accessions worldwide, is by far the best-collected root crop, both in terms of landraces and wild species. Virtually all potato landraces and 40 percent of the wild species are now in gene banks (Table 6.1). The relative completeness of these collections is largely due to the leading role of the Lima-based International Potato Center (CIP—Centro Internacional de la Papa) and the keen interest in many developing countries where the crop is grown (Figure 6.3).

CIP holds close to a third of the world potato-germplasm collection (Table 6.9). Of the 6,500 potato accessions maintained there, 5,000 are clones that are grown each year near Huancayo in the Andes (Figures 6.4, 6.5). Duplicates are sent to CIP's headquarters in Lima and to the Colombian national program (ICA—Instituto Colombiano Agropecuario) in Bogotá. The potato was domesticated in Peru, and that country accounts for 82 percent of CIP's accessions (Huamán, 1982). About 80 percent of CIP's collection consists of the common or Irish potato (*Solanum tuberosum*), but more restricted domesticates, such as *S. ajanhuiri* and *S. stenotomum*, are also represented. The center has 1,500 accessions representing ninety species of *Solanum* stored as seed under medium- and long-term conditions. Duplicates of seeds from wild potatoes are sent to the Inter-Regional Potato Introduction Project at Sturgeon Bay, Wisconsin, which houses the largest collection of wild potato seeds, with over ninety species from more than fifty countries (Hanneman, 1976).

Efforts are being made to develop true seed of potato for easier distribution to farmers and to facilitate long-term storage of germplasm.

TABLE 6.8
Pulse crops in gene banks.

Species/ Accessions	Type of Storage	Institution	Location
SOYBEAN (*Glycine max*)			
10,200	M#	AVRDC	Shanhua, Taiwan, China
8,350	L	NSSL*	Fort Collins, Colo., USA
3,500	M, L	NIAS	Tsukuba, Japan
3,000	M, L	VIR	Leningrad, USSR
3,000	S	OBCI	Wuhan, China
2,900	S	SAA	Shadong, China
2,900	M	CRIFC	Sukamandi, Indonesia
2,337	S#	NBPGR	Akola, India
1,500	M	INIA	Chapingo, Mexico
1,359	M, L	IITA	Ibadan, Nigeria
1,339	M, L	IPB	Los Baños, Philippines
1,000	S	LAA	Harbin, China
1,000	L	PGRO	Ottawa, Canada
COMMON BEAN (*Phaseolus vulgaris*)			
30,790	S, L	CIAT*	Cali, Colombia
9,321	S	WRPIS	Pullman, Wash., USA
8,900	M#	INIA	Chapingo,Mexico
5,000	M	CU	Cambridge, UK
4,456	L	NSSL	Fort Collins, Colo., USA
4,202	S	EMBRAPA	Goias, Brazil
3,851	M, L	ZGK	Gatersleben, German D.R.
3,109	S	ICA	Bogotá, Colombia
2,679	S	INIPA	Lima, Peru
2,627	S, L	NIAVT	Tapioszele, Hungary
2,575	S	NRPIS	Geneva, N.Y., USA
2,420	L	IPIGR	Sadovo, Bulgaria
2,000	S	NU	Nairobi, Kenya
2,000	S	MU	Lilongwe, Malawi
1,800	L	CENARGEN	Brasília, Brazil
1,712	S	THRS	Thike, Kenya
1,700	S	IIHR	Bangalore, India
1,500	S	UNA	Lima, Peru
1,242	S	ICCPT	Fundulea, Romania
1,162	S	FONAIAP	Caracas, Venezuela
1,042	L	INIA	Madrid, Spain
1,000	L	IVT	Wageningen, Netherlands
1,000	S	NIAR	Butare, Rwanda
LIMA BEAN (*Phaseolus lunatus*)			
3,846	M	NBI	Bogor, Indonesia
2,527	S, L	CIAT*	Cali, Colombia

TABLE 6.8 (*cont.*)

Species/ Accessions	Type of Storage	Institution	Location
RUNNER BEAN (*Phaseolus coccineus*)			
1,179	S, L	CIAT*	Cali, Colombia
1,500	M#	INIA	Chapingo, Mexico
PEA (*Pisum sativum*)			
4,090	L	IG	Bari, Italy
3,062	L	NGB	Lund, Sweden
3,058	M#	ICARDA	Aleppo, Syria
2,694	S	NRPIS	Geneva, N.Y., USA
2,067	M	JII	Norwich, UK
1,818	L	ZGK	Gatersleben, German D.R.
1,429	L	NSSL	Fort Collins, Colo., USA
1,296	L	FAL	Braunschweig, F.R. Germany
1,223	L	NIAVT	Tapioszelle, Hungary
1,011	L	PGRC	Addis Ababa, Ethiopia
GROUNDNUT (*Arachis hypogaea*)			
11,448	M, L	ICRISAT*	Hyderabad, India
6,299	S	NRCG	Junagadh, India
5,912	S	SRPIS	Experiment, Ga., USA
4,229	L	NSSL	Fort Collins, Colo., USA
4,000	M	NCSU	Raleigh, N.C., USA
4,000	M	TAMU	Stephenville, Tex., USA
4,000	M	OSU	Stillwater, Okla., USA
2,000	L	INTA	Manfredi, Argentina
1,800	L	IAC	Campinas, Brazil
1,800	M	VPI	Suffolk, Va., USA
1,730	M	CRIFC	Sukamandi, Indonesia
1,500	M	UF	Marianna, Fla., USA
1,461	S	ISRA	Bambey, Senegal
1,200	M	UF	Gainesville, Fla., USA
1,053	M, L	VIR	Leningrad, USSR
BAMBARA GROUNDNUT (*Vigna subterranea*)			
1,200	M, L	IITA	Ibadan, Nigeria
MUNGBEAN (*Vigna radiata*)			
5,736	M, L	UPLB	Los Baños, Philippines
5,483	M	AVRDC	Shanhua, Taiwan, China
3,000	S	PAU	Ludhiana, India
2,172	M	CRIFC	Sukamandi, Indonesia
2,100	S	UM	Columbia, USA
1,850	S#	NBPGR	New Delhi, India

TABLE 6.8 (*cont.*)

Species/ Accessions	Type of Storage	Institution	Location
COWPEA (*Vigna unguiculata*)			
12,000	M, L	IITA*	Ibadan, Nigeria
3,930	M	NBI	Bogor, Indonesia
2,537	M	CENARGEN	Brasília, Brazil
1,616	S	NBPGR	New Delhi, India
1,386	M, L	IPB	Los Baños, Philippines
1,150	S	SRPIS	Experiment, Ga., USA
1,050	M, L	VIR	Leningrad, USSR
CHICKPEA (*Cicer arietinum*)			
13,819	M, L	ICRISAT*	Hyderabad, India
5,585	M#	ICARDA	Aleppo, Syria
3,396	S	WRPIS	Pullman, Wash., USA
2,031	L	NSSL	Fort Collins, Colo., USA
2,000	S	IARI	New Delhi, India
1,685	M, L	VIR	Leningrad, USSR
1,600	M	INIA	Chapingo, Mexico
PIGEONPEA (*Cajanus cajan*)			
10,104	M, L	ICRISAT*	Hyderabad, India
1,500	S	IARI	New Delhi, India
LENTIL (*Lens culinaris*)			
5,906	M#	ICARDA	Aleppo, Syria
1,197	S	IARI	New Delhi, India
FABA BEAN (*Vicia faba*)			
3,293	M#	ICARDA	Aleppo, Syria
2,525	L	VIR	Leningrad, USSR
1,469	M, L	IG	Bari, Italy
1,136	L	FAL	Braunschweig, F.R. Germany
WINGED BEAN (*Psophocarpus tetragonolobus*)			
3,809	M	NBI	Bogor, Indonesia
1,000	S	IARI	New Delhi, India
1,000	L	TISTR*	Bangkok, Thailand

Note: See notes to Table 6.2.

* IBPGR-designed base collection.   # Long-term facility under construction.

Japan, for example, is assisting CIP in the construction of a long-term storage facility for true potato seed. This approach will not be feasible for all varieties of root crops, however. Some tuber crop varieties do not normally set seed, and saving seed of those that do does not maintain the integrity of the clone because of segregation of the seedling progeny.

Several agricultural research centers are using tissue-culture techniques to maintain clean root-crop germplasm. At CIP, for example, growth of potatoes in test tubes is slowed by cool temperatures and the use of certain culture media; at 6-10° C, potato plantlets thrive in test tubes for two years. CIP still maintains the bulk of its landrace material as clones in field plantings, and these are harvested annually; but tissue culture promises to save considerable space and time. At CIAT, the conversion of cassava germplasm to tissue culture is further along. Two-thirds of the center's 3,700 cassava accessions have been cultured as plantlets in test tubes, and CIAT is duplicating its entire field collection in tissue-culture form. This technique is also being used by IITA to maintain cassava, sweet potato, cocoyam (*Xanthosoma* spp.), yams, and taro, and by the Asian Vegetable Research and Development Center in Taiwan to maintain cultures of sweet potato.

Germplasm conservation of cassava, yams, and sweet potato, important staples in the humid tropics, is not nearly as advanced as that of potato. This relative lack of progress partly reflects the absence of these crops in large-scale, interregional trade. Only 35 to 50 percent of the landraces of cassava, sweet potato, and yams, and 5 percent or less of the relatives of these crops, are in gene banks (Table 6.1). Sizable collecting gaps thus exist for landraces and wild relatives of most root crops.

In the case of cassava, for example, numerous traditional varieties still await collection, and CIAT's cassava germplasm collection is expected to double to accommodate landraces and duplicates from other collections (CIAT, 1985:37). One reason for this large number is the diversity of ethnic groups that grow the crop, with each tribe or group often having a unique repertoire of landraces.

Only the surface of Amazonia's wealth of cassava genotypes, for example, has been scratched. In the Uaupés watershed of northwest Amazonia, Janet Chernela, an anthropologist, has recorded 135 distinct cassava cultivars in four villages of the Tucano (Chernela, in press). The Jívaro of the Ecuadorian Amazon cultivate over 100 cassava varieties (Boster, 1983); the Kuikuru, who inhabit the upper Xingu watershed, recognize some 50 distinct cassava cultivars (Carneiro, 1983); and the Desana of northwest Amazonia maintain at least 40.

Northwest Amazonia is a particularly promising area for cassava-germplasm collecting missions, partly because of the relative isolation of Indian groups, who maintain many of their traditional ways. But other parts of Amazonia still inhabited by aborigines also contain rich germplasm pockets of cassava and other crops. For example, one group

6.3. Several varieties of potato for sale in Huancayo, Peru,
June 1982.

of the Kayapó who live between the Xingu and Tocantins rivers raises
an enormous diversity of cassava in the forested region. Warwick Kerr,
a Brazilian geneticist, and Darrell Posey, an American anthropologist,
found twenty-two cultivars of cassava in Kayapó fields and collected
them for Brazil's program for the conservation of crop genetic resources (Kerr and Posey, 1984). More wild cassavas need to be collected
from such centers of diversity in northern Brazil and southwest Mexico
(Hawkes, 1985:40).

Collecting the germplasm of root crops of local importance, such as
ullucu (Figure 6.6) of the high Andes, was started by IBPGR in the late
1970s. Ullucu is a food crop, but its tremendous variation in shape and
color is due primarily to selection for medicinal and magical properties.

The thumb-sized tubers are thought to help with childbirth and internal injuries (Patiño, 1963).

Breeding work for most tropical root crops other than potatoes is relatively recent, and few high-yielding varieties have been released to farmers. Erosion of landraces of cassava, sweet potato, yams, and most other root crops is thus not as serious as it is for the cereal crops. But as tribes become extinct and farmers change their cultivation patterns, landraces are constantly being lost, and efforts to conserve the genetic diversity of root crops will likely continue into the next century.

VEGETABLES

Vegetables are, by and large, better collected than root crops, partly because they are easier to store as seed. In 1980, IBPGR initiated a major effort to conserve vegetable crop germplasm that coincided with similar efforts by various national and regional programs. This major push to preserve the genetic diversity of vegetables resulted in the doubling in size of vegetable germplasm collections within five years. Sizable collections of numerous vegetables are now in place and, in most cases, significant diversity of the landrace material has been gathered (Table 6.1). And since IBPGR has played a major catalytic role in the germplasm collection of vegetables, the situation with regard to the preservation of wild species is better than with cereals and grain legumes. The collection of close relatives of vegetable crops will accelerate further and is expected to be reasonably complete by 1990, although many species in secondary gene pools will continue to merit attention.

Among vegetable crops, tomato and its wild relatives have been more fully collected (90 and 70 percent, respectively), and the *Capsicum* peppers are next best represented in gene banks. Virtually all the primitive cultivars of amaranths have been collected. Considering the value of horticulture in industrial countries and the growing importance of vegetables in the tropics, particularly to supply burgeoning urban populations (Figure 6.7), the increased attention to conserving vegetable germplasm is justified. For many people, vegetables are the main source of vitamins, minerals, and protein.

INDUSTRIAL AND FORAGE CROPS

The germplasm conservation of some industrial crops is a little over the half-way point (Table 6.1). Of the industrial crops, sugar cane and cotton collections are most numerous, though generally modest (Table 6.10). As is to be expected these collections are concentrated in coun-

TABLE 6.9
Root and tuber crops in gene banks.

Species/ Accessions	Type of Storage	Institution	Location
POTATO (*Solanum* spp.)			
9,435	M, L	VIR	Leningrad, USSR
6,500	S, M. L	CIP*	Lima Peru
5,000	S, M	CENARGEN	Brasília, Brazil
4,286	S	UB	Birmingham, UK
4,000	S	INIA	Toluca, Mexico
2,800	M	IRPIS	Sturgeon Bay, Wisc., USA
2,605	M	CPRI	Simla, India
2,370	S, M	FAL**	Braunschweig, F.R. Germany
1,282	S	AVRDC	Shanhua, Taiwan, China
1,250	M	SCRI***	Roslin, UK
1,000	S	ICA	Tibaitata, Colombia
SWEET POTATO (*Ipomoea batatas*)			
1,243	S	CIP	Lima Peru
1,200	S	NBI	Bogor, Indonesia
1,200	S	KNAES	Kagoshima, Japan
1,200	S	AVRDC	Shanhua, Taiwan, China
1,000	S	IITA	Ibadan, Nigeria
CASSAVA (*Manihot esculenta*)			
3,700	S	CIAT*	Cali, Colombia
1,829	S	IITA	Ibadan, Nigeria
1,327	S	CTCRI	Kerala, India

Note: See notes to Table 6.2.
* IBPGR-designated base collection for seeds.
** Houses the Dutch-German potato gene bank.
*** Houses the historically significant Commonwealth Potato Collection.

tries where the crops are important for domestication use and for export. Cacao holdings, for example, are largest in West Africa and Trinidad, where the crop provides valuable foreign exchange, and India has a large share of the world's assembled cotton germplasm.

Only three germplasm collections of rubber exceed one thousand accessions, and only one field gene bank for African oil palm has more than one thousand samples. By virtue of the difficulties of growing large numbers of palms, the germplasm of coconuts is inadequately represented in collections. Banana collections are unlikely to need to grow beyond one thousand accessions.

The relative lack of progress in preserving the genetic diversity of industrial crops can be explained by several factors. First, many of the seeds of industrial or plantation crops, such as rubber and cacao, remain viable for very short periods and cannot be stored in conventional gene banks because they are damaged by drying and freezing. Second, the international agricultural research centers are concerned with food crops, so their leadership, scientific expertise, and sustained funding have not been applied to the export-oriented cash crops. Third, the job of conserving the germplasm of such crops has been largely left in the hands of private companies and a few national programs. Short-term breeding objectives have generally governed the germplasm-acquisition efforts of such organizations. Reliable support for sizable germplasm collections is difficult for companies concerned with cutting costs, or for national programs holding the germplasm in the wild when countries in other parts of the world use the materials. Action by the public sector is needed to provide incentives for the private sector to contribute to germplasm conservation.

With IBPGR input, modest progress in germplasm conservation has been made with several industrial crops, particularly beet, cacao, and cotton. Other efforts to collect industrial crop germplasm are underway, but industry is generally expected to provide much of the funding for future genetic resources work.

OVERALL COLLECTION AND EVALUATION GAPS

Although some 130,000 accessions of legumes and 85,000 samples of grasses are in forage collections worldwide, little is known about the degree of overlap and coverage of the entries. In spite of the impressive numbers, germplasm conservation is still largely in its infancy and major collecting efforts will commence in the near future. Arid areas, the Mediterranean region, and the tropics and subtropics, will be the focus of much of the future forage germplasm collecting. To further this task, IBPGR is arranging partnership agreements with some sister centers within the Consultative Group on International Agricultural Research (CGIAR), such as the International Livestock Center for Africa (ILCA), ICARDA, and CIAT, as well as national programs, especially in Australia and Brazil.

In order to identify collecting omissions in the world gene pool of a crop, the genetic diversity of existing collections must be ascertained. Such knowledge becomes available only after the passport data on accessions have been analyzed and the samples have been evaluated. Unfortunately, this type of information is often lacking at gene banks. Furthermore, plant collectors need to have an idea of what remains in

6.4. A sample of the genetic diversity of potatoes in the germplasm collection of the International Potato Center, Lima, Peru.

the field, particularly in centers of diversity. Ecogeographical survey work and taxonomic studies are crucial for this background information.

Significant progress has been made in sampling the genetic diversity of many crops, but many gaps still remain for wild species. Future emphasis in collecting work should thus be placed on crop relatives. Large unexplored areas remain for some crops, especially for species of local or regional importance. Collecting gaps are also large for many underused or unexploited plant species that could play a more important role in commerce and subsistence. In some instances, emergency situations arise that warrant immediate attention; the Sahel drought, which is eliminating the genetic diversity of crop landraces and wild species, is such a case.

IBPGR is now stressing the importance of collecting landraces in iso-

136

6.5. Cross section of several varieties of potato with different color patterns in the germplasm collection of the International Potato Center, Lima, Peru.

lated areas and assembling a richer germplasm collection of crop relatives. Germplasm collections of wild species will need to be manageable to keep costs down and to ensure adequate care and evaluation. The goal in collecting close relatives of crops should be to safeguard threatened populations and to gather sufficient variability to meet the requirements of plant breeders now and in the foreseeable future. For example, the collection of an additional 1,000 populations of wild *Triticum* and 500 genetically distinct samples of *Aegilops* is warranted to bolster wheat collections in gene banks (Chapman, 1984).

THE GROWTH AND LOCATION OF GENE BANKS

Until relatively recently, most gene-bank collections were small and concentrated in the industrial countries. In 1976, for example, only a

6.6. Harvesting ullucu (*Ullucus tuberosus*) in the
Mantaro Valley, Peru, June 1982.

handful of seed stores for preserving crop genetic diversity were in
place, and most of them were in developed nations. Within the last decade, however, the number of gene banks in the Third World has risen
dramatically. In 1984, a total of 113 seed storage facilities in fifty-eight
countries had been built; of these, 43 were designed by IBPGR for long-
term storage. By 1985, at least seventy-two countries, approximately
half the total number of nations, had germplasm facilities operating or
under construction (Appendix 3). Eighteen countries are now building
gene banks that are expected to be ready in 1986.

Since its inception, IBPGR has channeled most of its efforts towards
collecting the genetic diversity of the major crops and the building-up
of a network of base collections so that the germplasm is secure for the
future. The Board adopted a pragmatic approach to assembling the
network by using existing institutions that were capable of long-term
seed storage, while helping others gain that capacity, especially in developing countries. The number of gene banks with base collections
within the IBPGR global network grew from five in 1976, to thirty in
1983, and thirty-three by March 1985 (Appendix 4). Base collections
within the IBPGR network are held in over two dozen countries and
represent thirty-four staple seed crops (Hanson et al., 1984:2). IBPGR
has set a goal of fifty base seed collections covering forty major crops in
its network. With the addition of field gene banks (Appendix 5), the
present number of gene banks designated as security collections increases to forty-six, twenty-nine of which are located in the Third World.

A major future task is to build up a network of gene banks with me-

6.7. Assorted vagetables and fruits for sale in Bangkok,
Thailand, 1971.

dium-term collections in order to supplement the network of base collections. Over one hundred such collections are significant enough to
warrant tighter coordination. This complementary network of medium-term collections is needed to improve and streamline the multiplication, regeneration, characterization and evaluation, documentation, and exchange of germplasm accessions. IBPGR will accelerate its
efforts in this direction over the coming years.

Closer communication between gene-bank operators is necessary because many gene banks are currently operating inadequately, especially
with respect to the loss of materials due to such factors as insufficient
drying of seeds prior to storage, the incorporation of diseased material,
and technical refrigeration problems. In some cases, funding for the
monitoring and maintenance of base seed collections is meager. Although the amount of genetic resources in gene banks has increased
considerably, many gene bank samples could be lost due to poor maintenance.

A recently published summary of the last ten years of collecting activity listed about 85,000 accessions spanning 138 crop species. The acces-

139

TABLE 6.10
Industrial crops in germplasm collections.

Species/ Accessions	Type of Storage Form	Institution	Location
SUGAR CANE (*Saccharum* spp.)			
3,533	clones	FSC	Lautoka, Fiji
3,400	clones	SFS	Canal Point, Fla., USA
3,000	clones	SIRI	Mandeville, Jamaica
2,713	clones	SBI	Coimbatore, India
2,000	clones	HSPA	Maunawili, Hawaii, USA
1,774	clones, seeds	AES	Macknade, Australia
1,600	clones, seeds	WICSCBS	St. George, Barbados
1,500	clones	IMPA	Balderas, Mexico
1,300	clones	CTC	Antonio, Brazil
1,259	clones	INICA	Havana, Cuba
1,233	clones, seeds	MSES	Gordonvale, Australia
CACAO (*Theobroma cacao*)			
6,000	trees	CRIG	Tafo-Akin, Ghana
3,700	trees	CRU	St. Augustine, Trinidad
TEA (*Camellia sinensis*)			
15,000	plants	TRIEA	Kericho, Kenya
1,761	plants	NRIT	Shizouka, Japan
1,000	plants	TRFCA	Malanje, Malawi
COFFEE (*Coffea* spp.)			
6,000	plants	IRCC	Divo, Ivory Coast
1,284	plants	IAR	Jimma, Ethiopia
1,212	plants	CATIE	Turrialba, Costa Rica
COTTON (*Gossypium* spp.)			
5,244	S	CICR	Nagpur, India
4,900	L	VIR	Leningrad, USSR
3,905	S	PAU	Ludhiana, India
2,518	S	CICR	Coimbatore, India
2,400	S	CGR	College Station, Tex., USA
2,000	L	IRCT	Montpellier, France
OIL PALM (*Elaeis guineensis*)			
1,300	plants	PORIM	Kuala Lumpur, Malaysia
RUBBER (*Hevea brasiliensis*)			
3,400	plants	IRCA	Abidjan, Ivory Coast
2,500	plants	RRIM	Serdang, Malaysia
1,000	plants	CNPSD	Manaus, Brazil

Note: See notes to Table 6.2.

sions were collected in eighty-five countries and are deposited in long-term storage facilities as well as in their countries of origin. Compared to a decade ago, the germplasm conservation situation is much more satisfactory because scientists, politicians, and the general public are more aware of the importance of collecting and conserving of crop germplasm. The steep rise in the number of gene banks is concrete evidence of the increasing realization among policy makers that it is imperative to preserve the genetic diversity of crops and their wild relatives for posterity.

GENE-BANK DIVIDENDS

In previous chapters we have outlined the rationale for gene banks, the way they operate, and the range and scope of germplasm collections. We pointed out that breeders need to be able to fall back on diverse and thoroughly evaluated germplasm accessions if they are to continue to help raise agricultural productivity and to prevent drastic yield fluctuations. One can advocate the preservation of plant genetic diversity on scientific, aesthetic, and even moral grounds, but, for better or worse, economics is also a powerful motivator of human behavior. The preservation of plant germplasm will be easier if it can be demonstrated that economic payoffs ensue from safeguarding the products of millions of years of evolution and thousands of years of human experimentation. And as the world's population continues to grow and ever-increasing demands are placed on Earth's resources, couching the cause of conservation in economic terms will enhance the chances of success.

Does it pay to have gene banks? In other words, are they cost effective? A dollars-and-cents answer to this question is not easy because the economic payoffs of improved varieties developed through the use of gene banks are rarely specific. But we can indicate roughly how much is spent worldwide on plant genetic resources and how germplasm collections have contributed to crop improvement. It will be evident that although gene-bank dividends are difficult to gauge in monetary terms, the balance sheet for gene banks is nevertheless overwhelmingly positive.

Approximately $55 million was spent worldwide on plant genetic resources work in 1982 (Table 7.1). This figure has not changed significantly since then. Indeed, the amount spent on gene bank operations in the mid-1980s has probably shrunk slightly since 1982 after adjustment for inflation. The $55 million total is rough because it is difficult to draw the line between germplasm evaluation and crop breeding. Nevertheless, it is a ballpark figure—and a modest one, considering the task at hand: the price of a new, twin-engined Boeing 767 is the same as the annual expenditure on gene banks worldwide. Unfortunately, many Third World countries often opt for "prestigious," showy projects such as jumbo jets rather than lay the groundwork for lasting economic development. Only about $1.25 million is spent annually on rice gene banks throughout the world—a small sum, considering that the cereal

TABLE 7.1
Estimated global funding for crop genetic resources work in 1982.

Institution(s)	Dollars, in Millions
National programs in developed countries	$ 28.98
National programs in developing countries	7.41
International agricultural research centers	9.13
IBPGR	3.79
Bilateral aid and foundations	3.00
Multilateral (mostly United Nations)	2.78
Total	$ 55.09

Source: Plucknett et al., 1983.

is the main food for half of humanity (Chang, 1984a). The U.S. National Plant Germplasm System's 1985 budget of $13.9 million was the same as the previous year (Murphy, 1985), and is a modest amount considering that U.S. farm exports typically earn between $30 and $45 billion every year.

Crop breeders draw on gene banks to raise yield potential, improve nutritional qualities, and tackle a broad range of challenges to agricultural productivity. In some cases, gene banks contain superior material that is suitable for release with no further improvement. Since 1979, for example, elite bean accessions collected in six countries and held in the CIAT (Centro Internacional de Agricultura Tropical) gene bank near Cali, Colombia, have been released directly in nine Latin American countries (CIAT, 1984b:17).

In most cases, though, the contribution of gene banks to crop improvement is more indirect. Crop breeders often organize their working materials into gene pools with shared major characteristics, such as tolerance to cold. Gene banks provide a backstopping service to breeding programs by regularly supplying new materials for such specialized breeding pools. When searching for a trait, breeders usually turn first to their working collections because the plants are agronomically superior and are thus closer to what farmers want. Only when the desired genes are not found in elite breeding lines do scientists turn to collections in medium- or long-term storage.

To increase the likelihood that desirable genes will be located in elite lines when needed, gene-bank accessions are constantly fed into breeders' gene pools. This constant enriching process sometimes makes it difficult to pinpoint the specific source of a desirable trait. At the International Maize and Wheat Improvement Center (CIMMYT—Centro

Internacional de Mejoramiento de Maiz y Trigo), for example, the maize-breeding program incorporates some accessions from its medium-term gene bank every year into the thirty-three gene pools used for breeding. ICA (Instituto Colombiano Agropecuario), the Colombian national agricultural research program, annually uses one-quarter of the one thousand accessions in its potato gene bank for crossing experiments. By reaching deeply into its potato collection, ICA has successfully released twenty-four improved varieties, some of which have proved suitable in Ecuador and Venezuela (Gomez, 1985). Improved potato varieties now occupy about 85 percent of the potato-growing area of Colombia.

In this chapter we focus on efforts to improve and sustain agricultural productivity and to adapt crops to unfavorable environments. We spotlight ways in which breeders use germplasm collections to upgrade the disease and pest resistance of crops and to tailor varieties to adverse soils and climates. Any one of the cases discussed is probably responsible for increased agricultural production in excess of $55 million. For the United States alone, the contribution of germplasm collections to increased agricultural production is worth about $1 billion annually (Myers, 1983:3, 29); furthermore, about half of the doubling of wheat and rice yields in the United States since 1930 is due to breeding.

DISEASE RESISTANCE

Crop diseases, more than any other environmental challenge, account for the bulk of breeders' requests for gene-bank accessions. Resistance to insect pests is the second most sought-after trait in germplasm collections. Diseases and pests top the list of accession demands because pathogens and insects are present wherever crops are grown, are frequently abundant, and are constantly evolving. They are particularly abundant in the tropics, where no cold season checks their development and where they evolve more rapidly than temperate pests and pathogens because of the year-round growing conditions. Because of the dynamic nature of pest and pathogen populations, gains in breeding for disease or insect resistance are usually only temporary. New sources of resistance are therefore continuously sought.

Numerous cases of the successful tapping of germplasm collections to launch disease-resistant cultivars can be cited. We shall select examples from different crops in various regions to illustrate the pervasive nature of pathogen and insect challenges to agriculture. And because the environmental challenges to agriculture are especially severe in the tropics, many of the cases cited will be from warmer climates.

Fungi are among the most common causes of crop diseases, and although fungicides can sometimes be helpful, they are often too costly and provide only temporary control. Genetic resistance to fungal diseases is thus the best control method. On the northern plains of India, for example, the Indian Agricultural Research Institute (IARI) released semidwarf barley varieties in 1974 with proven resistance to yellow rust (IARI, 1980). IARI has also developed a popular maize hybrid, Ganga 5, that resists stripe downy mildew (*Scelophthora rayssiae* var. *zea*), leaf blight, and stemborer (Singh, 1980). Scientists in the Department of Vegetable Crops and Floriculture at the Himachal Pradesh Agricultural University in India located resistance to pea blight (*Ascochyta pinodella*) in a landrace, Kinnauri, which they used in crosses with modern pea cultivars to reduce the considerable damage caused by the fungal disease (Rastogi and Saini, 1984).

Fungi often mutate into new races, triggering a turnover of cultivars. In the late 1940s, for example, race 15B of black stem rust of wheat arose in the wheat-growing areas of the United States and Canada. Since all wheat cultivars planted in North America at that time were susceptible to the newly evolved pathotype, wheat production was threatened. Breeders screened their working collections for sources of resistance. Fortunately, after checking numerous wheat lines, breeders identified a few strains in their nurseries that resisted the new variant of stem rust (Adams et al., 1971). Resistant material included elite breeding lines developed from temperate wheat germplasm and a wheat from Kenya, which underscores the value of a diverse germplasm base for breeders. While this resistance was being incorporated into commercial bread-wheat varieties, race 15B took its toll, and for several years during the 1950s, the mutated fungus nearly halted durum-wheat cultivation in the Northern Great Plains.

Bacterial diseases of crops, as in the case of fungal pathogens, are difficult and expensive to control by chemical applications. In many cases, such as citrus canker, infected plants have to be destroyed to eradicate the disease from an area. Such radical measures drive up prices for consumers, and they bankrupt farmers. Genetic resistance to bacterial pathogens is thus highly desirable.

Because of its sustained investment in the agricultural sciences, India has enjoyed some remarkable successes in developing crop varieties resistant to pathogenic bacteria. IARI, for example, developed one of the most widely adopted maize varieties in India's history, Ganga Safed 2, which resists bacterial rot and pythium stalk rot (Singh, 1980). In Mexico, efforts to expand cassava production in the early 1970s were stymied by cassava bacterial blight (*Xanthomonas manihotis*) and super elon-

145

gation (*Esinoe brasiliensis*), a fungus disease. So Mexico turned to CIAT, which located a cassava landrace, Sabanera, in its field gene bank that resists the bacterial pathogen and tolerates super elongation. A CIAT-organized team collected Sabanera in the municipality of Ocú in Panama in 1970. The accession was introduced to Mexico in 1977 and has greatly boosted cassava production in Tabasco (Melendez et al., 1981; C. Hershey, pers. comm.).

The International Potato Center (CIP—Centro Internacional de la Papa) has developed varieties of the common potato (*Solanum tuberosum*) resistant to bacterial wilt by turning to another cultivated potato (*S. phureja*) for resistance genes. Bacterial wilt is one of the most important limiting factors for potato production in the tropics, but varieties of the common potato resistant to the widespread disease are now being grown in Peru, Brazil, Fiji, Indonesia, Sri Lanka, Nepal, Nigeria, Kenya, and Egypt (CIP, 1984:68).

Viruses are another important group of pathogens of crop plants that not only lower yields but impede the exchange of germplasm between countries. New viral diseases are constantly being discovered, and breeders must then launch a survey of genetic resources to combat them. Resistance to viral diseases is found in elite breeding lines, landraces, or wild species. In the southern Corn Belt of the United States, for example, a new virus, chlorotic-dwarf maize virus, began attacking maize in the late 1970s. Resistance to the new disease was located in advanced breeding lines that traced part of their ancestry to a Cuban open-pollinated maize, the source of the resistance (Duvick, 1984).

Considerable time is saved when resistance genes are already in place in agronomically desirable breeding lines. In their search for resistance to another new virus disease of maize, corn lethal necrosis, breeders also found suitable material in their working stocks in a pair of elite inbred lines. Both resistant lines traced their ancestry to an open-pollinated variety from Argentina.

Resistance to virus X of potato was located in a landrace of the Chilote Indians of Chiloe Island, Chile; this resistance has been incorporated into cultivars in Scotland and the United States (Ross, 1979). Another landrace of the Chilote Indians was used as a parent in a potato line that has been used widely in the German Democratic Republic and the Soviet Union for resistance to virus Y and to potato wart (*Synchytrium endobioticum*).

PEST RESISTANCE

In the literature, pests sometimes include pathogens and insects that attack plants and animals. Here we confine pests to insects that consume

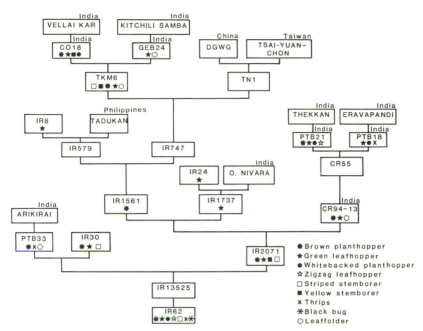

7.1. Sources of insect resistance in the pedigree of IR62, a rice variety developed by the International Rice Research Institute, Los Baños, Philippines. (From Heinrichs, 1984.)

plant tissue or fluids or transmit diseases when they feed on crops. Some insects, particularly members of the homopteran and hemipteran orders, damage crops by simultaneously removing fluids vital for growth and transmitting pathogens, especially viruses. Insect pests of crops range in size from microscopic mites to cigar-length locusts.

Gene banks are heavily used by breeders screening for material that withstands insect attack. In a survey of rice breeders at twenty-seven experiment stations in ten Asian countries, insect resistance was one of the objectives in 41 percent of the crosses (Heinrichs, 1984). After early lines of dwarf rice were seriously attacked by pests in the mid- to late 1960s, scientists at the International Rice Research Institute (IRRI) tapped the institute's gene bank to develop new rice lines resistant to several insect pests.[1] IR62, for example, released in the Philippines in 1984, resists more insects than its predecessors, including the widely planted IR36 (IR2071) (Figure 7.1). As early as 1973, only seven years after the release of IR8, which started the green revolution in rice, researchers in Vietnam and the Solomon Islands used IRRI material as

[1] For a profile of a green-revolution rice with resistance to multiple pests and diseases, see Chapter 9.

parents in crosses to develop brown planthopper-resistant varieties (Khush, 1977).

IRRI's gene bank has also supplied material to combat sogata leafhopper (*Sogatodes oryzicola*), a serious pest in Latin America and a vector for the damaging oja blanca virus. CIAT scientists located resistance to the pest in IRRI gene-bank accessions collected in irrigated rice-growing regions of Asia, a surprising discovery considering that Asian rices are not under selection pressures from sogata leafhopper. IRRI germplasm resistant to sogata leafhopper was incorporated into breeding lines at CIAT that have led to the release of forty-five varieties by Central and South American national programs (P. R. Jennings, pers. comm.). The yield advantage of the resistant cultivars is about 1.2 tons per hectare over the varieties they have replaced. High-yielding rice varieties containing IRRI and CIAT germplasm now occupy one quarter of the rice-growing area of Latin America, up from 14 percent in 1974 and 2 percent in 1971 (CIAT, 1984c:22).

In Africa, breeders at the Nigerian-based International Institute of Tropical Agriculture (IITA) located clones among the institute's cassava germplasm collection that resist cassava mealybug (*Phenacoccus manihoti*) and green spider mites (*Mononychellus* spp.). Cassava is a major food item in the diet of 200 million Africans in thirty-one countries. But the starchy root crop's contribution to the African diet, already deficient in calories in many areas, is threatened by green spider mites and cassava mealybug that were first recorded in Africa in 1971 and 1973, respectively. The pests, like their host, are native to South America and probably hitched a ride on imported cassava clones. Cassava mealybugs suck sap and inject toxin into the crop, causing the elongated leaves to shrivel (Figure 7.2). The curled and shrunken leaves allow competing weeds to proliferate and reduce the harvest of protein-rich leaves. The introduced pests spread rapidly and now thrive in half the cassava-growing region of Africa. Cassava mealybug and green spider mites have provoked an average 30 percent continent-wide dip in cassava production and are costing African farmers approximately $2 billion a year.

Scientists at IITA have adopted a two-pronged strategy to thwart the spread and damage caused by cassava mealybug and green spider mites. Several predators of cassava mealybug have been imported to Nigeria for testing, and initial results are promising (N. Smith, 1985). The other main thrust in the biocontrol effort is the development of cassava cultivars with genetic resistance to the imported pests. After screening cassava collections at IITA, clones have been identified that thrive in the presence of cassava mealybug and green spider mites (Leuschner,

7.2. Cassava field attacked by cassava mealybug (*Phenacoccus manihoti*) near Ibadan, Nigeria, May 1983.

1981, 1982). Pubescent leaves of the resistant clones discourage the pests from alighting (Figure 7.3). This morphological defense mechanism does not provide complete protection against the exotic pests, but when combined with the release of predators, it promises to shield cassava from significant damage. IITA has developed a high-yielding, resistant cassava line that is currently being multiplied by the Nigerian Seed Service for distribution to farmers.

When an accession has a useful gene, or set of genes, it cannot usually be immediately released to farmers. Much breeding work is normally required to transfer the desirable genes to agronomically suitable cultivars. Numerous backcrosses are often needed to shed unwanted characteristics while retaining the valuable traits. Furthermore, candidates for release usually undergo further testing by national seed certification organizations before distribution to farmers. At the gene bank operated by IITA, for example, a single source of resistance to a bruchid beetle (*Callosobruchus maculatus*), a widespread pest that tunnels into legume seeds, was found in 1975 after screening seven thousand cowpea

7.3. Cassava, with pubescent leaves, that resists cassava mealybug found in the germplasm collection at the International Institute of Tropical Agriculture, Ibadan, Nigeria, 1983.

accessions. It then took eight years to incorporate bruchid beetle resistance into seven agronomically desirable lines that are now being evaluated by national programs (IITA, 1983:49).

TOLERANCE TO ADVERSE SOILS AND CLIMATES

Tailoring crops to problem soils is another high priority of breeders. For instance, researchers at CIMMYT and the Brazilian national agricultural research program (EMBRAPA—Empresa Brasileira de Pesquisa Agropecuária) are screening germplasm collections for bread wheats and triticales (wheat and rye hybrids) that perform well on the acidic, high aluminum-content soils of the cerrado region of central Brazil.[2] Roots are stunted when soils contain toxic levels of aluminum,

[2] Triticale is an interspecific cross between wheat and rye. It is used for livestock feed and for human food in both developing and industrial countries (N. Smith, 1983b).

thus rendering the plants more vulnerable to drought. A CIMMYT triticale was approved for release in the Brazilian cerrado in 1985, and when dwarf wheats from Mexico are successfully crossed with Brazilian landraces to develop new, high-yielding cultivars, the sparsely settled region may become an important food producer (Silva, 1976). Such a transformation would have the added bonus of taking some of the pressure exerted by settlers off the Amazon rain forest, probably the world's richest storehouse of unevaluated plant diversity. CIP, CIAT, IRRI, and IITA are developing lines of potato, cassava, forage plants, rice, and cowpea that thrive on rain-fed, acid soils with high levels of aluminum (CIAT, 1980). Tolerance to salinity in rice has been derived primarily from traditional varieties from southern India and Sri Lanka (Chang et al., 1982).

Crop breeders also screen germplasm for tolerance to adverse climatic conditions, such as too much or too little water, or temperature extremes. Japanese scientists found that samples of the Silewah rice variety gathered in the hills of Sumatra in 1974 and stored at the IRRI gene bank are more cold tolerant than cultivars from Hokkaido, even though Silewah comes from an altitude of 1,300 meters in the tropics (Chang, 1985). As in the case of sogata leafhopper resistance in Asian rices, plant populations sometimes contain genes of no current value. Presumably such "unused" genes are maintained to help ensure the long-term survival of a species in the event of environmental change, or they have a secondary unrecognized value in their native habitat. This occurrence of genes conferring resistance or tolerance to pests, diseases, and other adverse conditions away from areas of selection pressure is appreciated and exploited by scientists seeking material for gene banks and breeding programs.

To give the reader some idea of screening procedures for plant materials that survive cold temperatures, we will briefly describe the manner in which the Silewah variety was uncovered. First, scientists drew up a list of 729 candidates in the IRRI gene bank that had some promise of cold tolerance based on observations and collection areas. Seedlings of the 729 accessions were kept in water at 12° C for ten days (Figure 7.4); this hurdle eliminated 685 candidates (Vergara, 1984). The remaining 44 candidates were subjected to 15° C for five consecutive nights at the flower-initiation stage; this step reduced the number of promising candidates to 26. The surviving 26 samples were kept for one day at a diurnal temperature of 21° C during the anther-formation stage, and 15 candidates were then dropped from further screening. Japanese scientists tested six of the remaining candidates at different panicle devel-

7.4. Screening rice lines for cold tolerance using frigid water at the International Rice Research Institute, Los Baños, Philippines, 1985.

opment phases at 12° C for four days and only two accessions proved highly cold tolerant, one of which was Silewah.

Insufficient or excessive water can be detrimental to crops, and breeders have turned to germplasm collections in search of tolerance to those environmental stresses. Drought tolerance concerns rice breeders who are developing high-yielding varieties for upland areas. IRRI's gene bank is again proving to be a valuable resource in this breeding area since the rice-evaluation program of the Institute has uncovered 2,781 accessions that do well in dry areas (Chang, 1980). Collections made in Bangladesh during a flood in 1974 turned up rices that survive in water 5 m deep. IRRI and the Indonesian and Thai national programs have started deep-water rice breeding projects that tap traditional rices for tolerance to flash floods as well as to prolonged periods in deep water.

Deep-water rice research projects have been prompted by the sizable rice area subject to uncontrolled flooding. In southern Kalimantan, Indonesia, alone, swamps occupy some 2 million hectares (Noorsyamsi and Hidayat, 1974). In this low-lying, equatorial region, farmers must transplant rice three times during the rainy season to keep ahead of rising waters. The final transplanting, sometimes accompanied by a trimming of the leaves to promote tillering, occurs at the end of the rains, and the crop is harvested after a nine-month growing season. Although the several thousand landraces of the region are reliable performers, they produce only one to two tons per hectare and permit only one crop per year.

Suhaimi Sulaiman, a native of the region and a rice breeder at the Banjarmasin Research Institute for Food Crops in southern Kalimantan, accordingly set out to develop higher-yielding rices with a shorter growth period that would resist the prevailing diseases and pests and be acceptable to local tastes. Sulaiman, who received his master's degree in agricultural sciences at the Bogor Agricultural College in Java, targeted his breeding program for the relatively shallow swampy areas where he felt he could make an impact in the short term. Within a few years, Sulaiman's program, based on germplasm from Indonesia, Thailand, IRRI, and India, resulted in the release of two high-yielding rice varieties for the water-logged region. Mahakam, released in 1983, reaches 1.2 meters, contains germplasm from Thailand, Indonesia, and IRRI, yields three tons per hectare, and matures within 135 to 150 days. This shorter growth period creates the option of planting another crop. Mahakam does have one drawback; it is susceptible to brown planthopper. But because pesticides are used sparingly in Kalimantan, natural predators and a mosaic of traditional varieties keep brown planthoppers in check much of the time. Mahakam can thus be grown profitably in some areas.

Kapuas, the second variety developed by Sulaiman, builds on the strengths of Mahakam while eliminating some of its weaknesses, a typical pattern in varietal development. Kapuas is about the same height as Mahakam, contains Indonesian, Indian, and IRRI material in its pedigree, can tolerate submergence and peaty soils, resists biotype 1 of brown planthopper, and is moderately resistant to race 2 of the pest. Kapuas is also moderately resistant to brown spot disease, and yields the same as Mahakam but matures in 127 days, significantly earlier than its predecessor. Furthermore, Kapuas is palatable to locals. Much of the increased yield of Mahakam and Kapuas is due to genetic manipulation rather than changes in agronomic practices. Very little fertilizer is applied to the new high-yielding varieties in Kalimantan because it is quickly washed away.

Mahakam and Kapuas accounted for only about 5 percent of the area planted to rice in swampy areas of Kalimantan in January 1985, but the area planted to the new varieties was increasing. In 1986 Kapuas performed well in trials in swampy areas of eastern Sumatra and is likely to be adopted there. As is to be expected, the new varieties were particularly evident near the experiment stations of the Banjarmasin Research Institute for Food Crops due to the demonstration effect. Farmers take particular care of the newer high-yielding varieties by erecting scarecrows of flapping plastic strips and building plastic fences around fields to keep out rats.[3] No such precautions were observed in fields planted to traditional varieties.

A brief profile of the working conditions and plans of Sulaiman will illustrate problems, concerns, and aspirations of many scientists in the Third World working on conserving and utilizing crop germplasm. Although Sulaiman is fortunate enough to work for a government that actively supports agricultural research, he still faces a number of obstacles in his work. His working collection of rices in Banjarmasin is kept in 20-liter kerosene cans in a room without air conditioning. Banjarmasin is close to sea level and has a year-round sultry climate; accordingly, as much as one-fifth of Sulaiman's material fails to germinate when planted. Money for an air conditioner is not the problem; the wildly fluctuating voltage in Banjarmasin, which regularly damages electrical equipment, is the culprit; voltage regulators do not always work if the current oscillations are too strong. In spite of these drawbacks, Sulaiman's work continues, and like many other Third World scientists he seeks further training abroad. It is such a combination of familiarity with local conditions coupled with the latest research methodologies that is required if developing countries are to be in a position to face the challenges of maintaining and using germplasm collections to help raise agricultural productivity.

[3] Plastic instead of metal fences, about a meter high, are used because metal fences rust quickly in the acidic waters of Kalimantan's swamps.

WILD SPECIES: THE WIDER GENE POOL

Wild relatives of crops have often played important roles in sustaining agricultural productivity. People have been adopting wild species for thousands of years, and after domestication, some of the resulting cultivars have continued to hybridize naturally with their wild relatives. Such spontaneous gene flow helps maintain the vigor of crops and can lead to the development of new crops. Plant breeders have turned to wild species mainly for sources of resistance to diseases, pests, and stressful environments, such as those with adverse climates. Wild species, which include plants in natural environments as well as weedy forms that thrive in disturbed habitats, are crucial to the never-ending effort to adapt crops to difficult environments and to confront challenges to agricultural production.

Over four decades ago, Vavilov (1940) called attention to the potential of crop relatives for improving agriculture. Wild species are especially good sources for resistance to diseases and pests as well as for tolerance to harsh growing conditions because they have had to thrive on their own (Harlan, 1984). Crops, on the other hand, have had the benefit of human protection, such as weed and pest control, and a more uniform environment for growth. Also, soil conditions have been made more suitable for crop plants through plowing, fertilizing, and irrigating. Wild plants have been under constant pressure from pathogens, pests, severe climates, and unfavorable soils and have evolved myriad strategies for survival. It is this arsenal of defensive traits, the product of millions of years of evolution, that is so valuable to agriculture. Without tapping relatives of crops for a variety of useful traits, plant breeders would be hard pressed in many instances to overcome yield constraints. Genetic resources of crops thus include gene pools of wild species as well as cultivated forms.

We begin this review of the role of wild plants in agricultural history by pointing out cases where weeds eventually became crops. Uninvited plants in fields are generally a nuisance since they compete with crops for nutrients and light. Weeds may be destroyed, left alone if they are too troublesome to deal with, or tolerated. Some weeds occasionally become crops if they are deemed sufficiently useful to merit saving seed or cuttings for planting. We will also discuss those unusual situations in which genes have spontaneously passed between crops and wild species, because this ancient genetic interchange has sometimes improved the hardiness of crops.

Although wild and cultivated plants still occasionally exchange genes when they grow close together, this important two-way flow of traits is diminishing due to grazing pressure around the borders of fields, the elimination of hedgerows to accommodate large machines and increase the cropping area, and the widespread use of herbicides. Fortunately, plant breeders are picking up where nature and the keen eyes of traditional farmers have left off. Here we explore ways in which scientists use wild species to upgrade disease and pest resistance in crops, to make cold-tolerant varieties, to improve their nutritional qualities, and to shorten them so that they are easier to harvest. Rather than list all the potentially useful attributes of wild species, we concentrate on interspecific crosses that have culminated in advanced breeding lines or in the release of cultivars.

Wild species are vital resources for breeders, but they are usually used as a last resort. Breeders are rarely familiar with close relatives of crops, are often misled by confused or incomplete taxonomies, and prefer to work with elite germplasm (Harlan, 1984). Breeders first turn to improved material in working collections when seeking specific traits. If a search through material of current interest fails to uncover the desirable character, they usually proceed to screen landraces. Only after exhaustively checking available germplasm of the cultivated species do breeders turn to wild plants.

Wild species generally have more undesirable characteristics than breeding lines or varieties; more work is thus required to shed the unwanted characteristics while transferring the valuable trait to an agronomically suitable crop (Hawkes, 1977b). Initial characterization and evaluation of wild species can also be especially time consuming and laborious. It can take up to fifteen years to introduce a valuable character from a wild species into a successful cultivar, and much longer when dealing with perennials. In the case of root knot nematode (*Meloidogyne incognita*), a pest of tomatoes and potatoes, resistance was first detected in a wild tomato (*Lycopersicon peruvianum*) in 1941. By 1944, a hybrid between this wild tomato and the cultivated species (*L. esculentum*) was achieved with some difficulty. But it was not until 1956, after fifteen years of backcrossing to the cultivated species, that the tenacious linkage with undesirable fruiting characteristics was finally broken (Rick, 1967). That prolonged effort, involving scientists at four institutions in Tennessee, California, Arkansas, and Hawaii, has led to the release of scores of tomato varieties with resistance to root knot nematode.

In the discussion of wild species in plant breeding, we briefly explore barriers to interspecific crossing and how they are overcome. Here we examine the use of bridge species in shunting genes from wild plants to cultivars, the selection of diploids with unreduced gametes (chromosome number not halved as in sexual cells) when attempting crosses with tetraploids, and techniques such as the doubling of chromosomes with chemicals and breaking chromosomes by irradiation. More sophisticated breeding methods that may prove valuable in the future, such as cell fusion, were covered in Chapter 5. Finally, we underscore the pressing need to broaden the scope of germplasm collections by including more wild species.

WEEDS AND CROP EVOLUTION

Weeds in the history of agriculture have been mostly a curse, but have also provided some blessings. Although the persistent invaders have reduced agricultural yields and increased the drudgery of farm life, they have occasionally contributed useful genes to crops and have sometimes been domesticated themselves. Weed-infested fields have served as recruiting grounds for new crops. Several important crops started as field weeds. Rye (*Secale cereale*) and oats developed from weeds in barley and wheat fields in the Middle East and in northern and western Europe. In Central America, two cultivated amaranths (*Amaranthus hypochondriacus, A. cruentus*) and chenopodium (*Chenopodium nuttalliae*) probably originated as weeds in cultivated plots (Wolf, 1959:53). Carl Sauer (1969:145) has suggested that amaranths, beans (*Phaseolus* spp.), squashes (*Cucurbita* spp.), and maize began as weeds in fields planted to cassava and sweet potato. And in West Africa, some cereals may have started as weeds in yam fields (Shaw, 1976).

Crops that started as weeds have passed through three main stages. At first they were pulled, cut, or dug into the soil. Then as weeds proliferated, some were tolerated because they were too troublesome to destroy, or because they provided useful products such as food or straw. The next step, planting, occurred when the weed had accompanied farmers long enough for them to find one or more uses for it and thus had become worthwhile for cultivation.

Weed-infested fields have served as primitive "laboratories" in the development of crops. In the case of cultivated potatoes (*Solanum* spp.), for example, weedy *Solanum* diploids have interbred with cultivated species to form new domesticates. *S. ajanhuiri*, a diploid potato cultivated at an elevation of 3,800 to 4,100 meters by the Aymara of southern Peru and northern Bolivia arose after a cultivated potato (*S. steno-*

157

tomum) and a wild species (*S. megistacrolobum*) hybridized (Huamán et al., 1980; Hawkes, 1981). The latter species is responsible for *S. ajan-huiri*'s frost tolerance.

The common potato (*Solanum tuberosum*) is a tetraploid (contains four sets of chromosomes) and resulted from a cross between a diploid cultivated species (*S. stenotomum*) and a weedy diploid (*S. sparsipilum*). The latter species imparted resistance to root knot nematodes, and the common potato adapts to a wider range of growing conditions than its parents. Resistance to more virulent pathotypes of a serious nematode pest, *Heterodera rostochiensis*, has also reached the common potato through widecrossing. Resistance to this destructive pest spread from a wild tetraploid (*S. oplocense*) to weedy *S. sucrense* and then to *S. tuberosum* (Hawkes, 1977b). Weedy potatoes in and around fields have thus contributed to the development of new potatoes and have upgraded resistance to pests and severe climates.

Weeds have also been a constant source of new genes for resistance to pests, diseases, and adverse weather in other crops. In the Middle East, for example, wheat, oat, and barley occasionally cross naturally with their wild relatives and exchange resistance genes (Zohary, 1970; Browning, 1974). Bread wheat is a cross between emmer wheat (*Triticum dicoccum*) and a wild grass of the genus *Aegilops*; the latter is responsible for bread wheat's cold tolerance (Feldman, 1976; Hawkes, 1980). Barley has benefited from a continuous introgression of genes from weedy relatives growing within and along the borders of fields (Baker, 1970b; Nevo et al., 1979). In East Africa, finger millet (*Eleusine coracana*) hybridizes with a related field weed, *E. indica* (Hussaini et al., 1977).

In Central America, teosinte and perhaps species of *Tripsacum*, wild relatives of maize, have been introgressing with maize for thousands of years. The weedy grasses account for some of the rich diversity and high productivity of maize in Mesoamerica (Galinat et al., 1956; Mangelsdorf et al., 1967; Wilkes, 1972, 1985). Teosinte is a close mimic of maize, and in Mexico and Guatemala grows along the borders of fields and within maize patches; only when the crop flowers is it possible to identify them in the field. Farmers traditionally weed out teosinte and other plants between furrows, but leave the mimic when it sprouts along planted rows. At Chalco, in the state of Mexico, teosinte is harvested along with maize since it serves as livestock feed. Some teosinte kernels pass through the animal whole and are reintroduced to fields in manure. In the Nobogame Valley of Mexico's Sierra Madre Occidental mountain range, farmers sense that teosinte "helps" maize, and they deliberately mix seeds of both species when planting (Wilkes, 1977b).

Chance crosses of wild and domesticated potatoes, wheat, rice, barley, finger millet, maize and many other crops have been spotted by farmers, and some of them have been saved for propagation.

Spontaneous exchanges of genes between crops and wild relatives still occur in fields, particularly in developing countries, but this enriching process is now less common. The heavy use of herbicides in modern agriculture, for example, has clearly boosted yields, but this widespread practice precludes natural gene flow between crops and wild species nearby. In Central America, heavy grazing pressure has eliminated populations of teosinte around the borders of maize fields in many areas; since 1900 the range of teosinte has been halved (Wilkes, 1972). While some evolutionary opportunities are being foreclosed due to changes in agricultural practices, breeders are increasingly acting as agents in the flow of useful genes from wild germplasm to cultivated species.

WILD SPECIES AND BREEDING

As in the case of accessions of cultivated species, the germplasm of wild species is mostly tapped to extract genes for disease and pest resistance (Frankel, 1977). In the case of viral pathogens of potatoes, for example, samples of a wild potato (*Solanum stoloniferum*) sent to the Max Planck Institute in the Federal Republic of Germany have been used to develop six cultivars with strong resistance to potato virus Y (Ross, 1979). Breeders have used the extensive collections of wild species at the Inter-Regional Potato Germplasm Collection at Sturgeon Bay, Wisconsin, to transfer immunity for virus X from *Solanum acaule*, and immunity to viruses A and Y from *S. stoloniferum* and *S. chacoense*, to cultivated potatoes (Creech, 1970). With over twenty wild species contributing genes to potatoes, the potato is one of the best examples of the importance of wild species in breeding.

Wild species have been helpful to breeders attempting to combat a viral disease of another root crop, cassava. Cassava mosaic disease is a widespread problem in Africa and parts of Asia; the only resistance found thus far has been located in a wild species (*Manihot dichotoma*) and in Ceará rubber (*M. glaziovii*), a wild species in South America that has been grown in limited areas in Africa and Southeast Asia for its latex, which is used to make an inferior rubber (Hahn et al., 1979, 1980a; Beck, 1982). Crosses between Ceará rubber and cassava started in East Africa in 1937 with some success, but little more was done on this project until scientists at the International Institute of Tropical Agriculture at Ibadan, Nigeria, revived cassava mosaic resistance breeding. Fortu-

159

nately, a line from one of the earlier interspecific crosses had been maintained in Nigeria, and IITA researchers have used this line in their resistance-breeding work. Resistance to the disease has now been incorporated into high-yielding lines that are being evaluated for release in India and several African countries (Hahn et al., 1980b; Terry and Hahn, 1982).

A wild oat, *Avena sterilis*, has provided resistance to Barley Yellow Dwarf Virus (BYDV) in oat varieties recently released in the United States and Canada (Comeau, 1984). The wild oat is now widely used in oat breeding. To prevent epidemics of BYDV in *A. abyssinica*, an oat cultivated in Ethiopia, a crossing program with a wild species, *A. barbata*, has been recommended (Comeau, 1984). The susceptibility of Ethiopian oat landraces to BYDV illustrates that traditional varieties may produce more stable yields than modern cultivars, albeit at lower levels, but they are not immune to disease and pest attack.

Wild species have served as a last resort for breeders screening germplasm for resistance to fungal diseases. As early as 1908, *Solanum demissum* was employed to transfer late blight resistance to the cultivated potato (Ross, 1979). Although this major gene resistance only lasted until 1936, when new races of late blight fungus evolved, *S. demissum* helped improve potato production for several decades. Even today, this wild potato provides minor genes that have imparted some resistance to the ubiquitous fungus in cultivated potatoes (Watson, 1970). The International Potato Center (CIP—Centro Internacional de la Papa), for example, has employed germplasm of *S. demissum* and landraces of the common potato to help launch numerous potato varieties in eighteen countries with varying degrees of resistance to late blight (Table 8.1).

Fungal resistance has also been transferred from wild species to plan-

TABLE 8.1

Sources of potato clones with resistance to late blight introduced by the International Potato Center (CIP) to developing countries.

Country	Clone and Source	Major Trait(s)
Burundi	Sangema (Mexico)	LB
Colombia	ICA Sirena (Wisconsin)	LB
Costa Rica	BR 63.65 (Wisconsin)	LB, BW, PLRV
	B-71-240.2 (Argentina)	LB, PLRV
Ecuador	INIAP-Bastidas (CIP)	LB
El Salvador	India 830 (India)	LB
Ethiopia	Anita (Mexico)	LB
Fiji	Domoni (Wisconsin)	LB, BW, PLRV

TABLE 8.1 (*cont.*)

Country	Clone and Source	Major Trait(s)
Kenya	BR 63.76 (Wisconsin)	LB, BW, PVY
	I-1035 (India)	LB
	CGN-69.1 (Sweden)	LB
	ASN-69.1 (Mexico)	LB
	America (Wisconsin)	LB, BW
Malagasy	CFK-69.1 (Mexico)	LB
	BR 63.76 (Wisconsin)	LB, BW
Malawi	Rosita (Mexico)	LB
	CFK-69.1 (Mexico)	LB
	BR 63.76 (Wisconsin)	LB, BW
Nepal	CFJ-69.1 (Mexico)	LB, Wart
	CFM-69.1 (Mexico)	LB, Wart
	NPI-106 (Germany)	LB, Wart, PLRV, PVY
	NPI-108 (Germany	LB, Wart, PLRV, PVY
Nigeria	BR 63.76 (Wisconsin)	LB, BW
Peru	Molinera (Wisconsin)	LB, BW, PLRV
	Caxamarca (Wisconsin)	LB, BW, PLRV
	Amapola (Wisconsin)	LB, BW
	Perricholi (CIP)	LB
Rwanda	Gahinda (Mexico	LB
	Kinigi (CIP)	LB
	Nseko (Mexico)	LB
	Gasore	LB
	Montsama (Mexico)	LB, Early
	Sangema (Mexico)	LB
Sri Lanka	Sita (India)	LB, HT
	Krushi (India)	LB, HT
Tanzania	CGN-69.1 (Sweden)	LB
	BR 63.76 (Wisconsin)	LB, BW
Vietnam	Dalat 004 (Mexico)	LB
	Dalat 006 (Argentina)	LB, PLRV
	Dalat 012 (Mexico)	LB
Zaire	Sangema (Mexico)	LB
	Montsama (Mexico)	LB, Early
	Kinigi (CIP)	LB
	Atzimba (Mexico)	LB
	Nseko (Mexico)	LB
	BR 63.76 (Wisconsin)	LB, BW

Source: CIP, 1984:66.
Abbreviations: LB = late blight resistance; BW = bacterial wilt resistance; PLRV = potato leaf roll virus resistance; PVY = potato virus Y resistance; HT = heat tolerance; Wart = wart resistance.

161

tation crops. A wild sugar cane (*Saccharum spontaneum*) collected in Java, for example, provided resistance to red rot in cultivated sugar cane, and thereby helped establish a thriving sugar industry in India (Martin, 1965; Prescott-Allen and Prescott-Allen, 1982b:62). Resistance to the highly destructive blue mold disease of tobacco has been located in a wild species, *Nicotiana debneyi* (Lucas, 1980). The first tobacco cultivars with resistance to blue mold, all derived from crosses with *N. debneyi*, were released in Europe in 1962. Unfortunately, blue mold-resistant cultivars are highly susceptible to other diseases and pests, so breeders are working on obtaining blue mold-resistant varieties that are immune to a broad range of pathogens.

Bacterial diseases have also prompted crop breeders to search wild germplasm for resistance genes. Ceará rubber, for example, the species used in resistance breeding for cassava mosaic disease, is also a source of genetic resistance to cassava bacterial blight (Hahn, 1978). And in Sudanese cotton plantations, resistance to bacterial blight is also due to a wild species, *Gossypium anomalum* (Prescott-Allen and Prescott-Allen, 1982b:61).

Breeders concerned with disease resistance not only draw on germplasm resources to tackle current difficulties but also to forestall potential problems. The dangers of producing hybrids based on a single source of cytoplasmic male sterility have been highlighted in the case of the 1970 outbreak of southern corn leaf blight in the United States. A similar vulnerability to disease has arisen with the widespread planting of hybrid sugar beet (*Beta vulgaris*), an important cash crop in many temperate countries. Only one source of cytoplasmic male sterility is currently employed in sugar beet hybrids, a tack that has led to a genetic narrowing of the crop's germplasm. Three new sources of cytoplasmic male sterility have been discovered in wild beets, however, and these wild beets are being crossed with sugar beet to supply new genes for this useful trait (Bosemark, 1979).

Pest resistance is also high on the priority list of breeders, and wild species have again proved helpful in a variety of crops. The tomato germplasm collection at the University of California at Davis, although relatively small by gene bank standards with approximately five hundred accessions, has aided breeders because stored material has been evaluated and because wild species are represented. One wild tomato (*Lycopersicon hirsutum*) from western Ecuador and Peru, for example, resists two species of spider mite, a leaf miner, greenhouse white fly, tobacco flea beetle, and the Colorado potato beetle (Rick, 1973). This tomato species possesses glands that emit a pungent odor that is thought to discourage pests (Rick, 1979). This virtual cornucopia of

8.1. Wild potato species (*Solanum* spp.) being screened for pest and disease resistance at the Huancayo substation of the International Potato Center, Peru, 1982.

useful genes in a single species is sure to prove helpful to tomato breeders in the future.

The potato is also plagued by pests, and wild species have been employed in breeding programs to combat them (Figure 8.1). In the case of worms that tunnel into the growing tubers, for example, a wild potato (*Solanum vernei*) has been crossed with the main cultivated species (*S. tuberosum*) to confer resistance to several pathotypes of *Globodera pallida*, a particularly important cyst nematode pest in the Andes and Europe (Ross, 1979). This wild potato has contributed resistance to eleven modern cultivars in the Netherlands and West Germany. Resistance to the most virulent races of nematodes (*Heterodera* spp.) is found only in wild potatoes (*S. vernei, S. famatinae, S. infundibuliforme, S. gourlayi*) (Hawkes, 1958, 1977a). Furthermore, nematode-resistant genes appear to be confined to wild potatoes from Northern Argentina, Bolivia, and Southern Peru.

Frost resistance in potatoes is chiefly derived from wild species, especially *Solanum acaule, S. megistacrolobum,* and *S. demissum* (Ross and

Rowe, 1969; Hawkes, 1983:95). Frost resistance in cultivated potatoes is a result of spontaneous crossing with wild species. Accessions of a wild potato (*Solanum acaule*) at the Sturgeon Bay Inter-Regional Potato Germplasm Collection have contributed cold tolerance to the common cultivated species (*S. tuberosum*) in addition to virus X resistance (Creech, 1970). Cold tolerance has also been successfully transferred from wild species of peppermint, grape, strawberry, wheat, rye, and onion to their cultivated relatives (Stalker, 1980). Heat- and drought-tolerant wild peas and wheat have contributed the responsible genes to their cultivated cousins, and wild tomatoes have contributed salt tolerance (Stalker, 1980; Harlan, 1984).

Wild species sometimes contain more protein than their cultivated forms (Harlan, 1984). The protein content of some varieties of cassava, for example, has increased as a result of crosses with two wild species, *Manihot melanobasis* and *M. tristis* (Jennings, 1976). Russian scientists have crossed winter wheat with a wild wheat grass (*Agropyron glaucum*) to improve the protein content of wheat grain and hay (Tsitsin and Lubimova, 1959). A diminutive wild tomato (*Lycopersicon pimpinellifolium*) has provided genes for higher vitamin content in the cultivated species (Rick, 1979). A wild cotton (*Gossypium thurberi*), native from Arizona to Mexico, has contributed fiber strength to upland cotton (*G. hirsutum*), the most widely grown species of the four cultivated cottons (L. L. Phillips, 1976). Attempts are being made to reduce the height of African oil palm by crossing the species, *Elaeis guineensis*, from West Africa with *E. oleifera*, a short, wild palm from lowland South America (Hardon, 1976). A wild rice (*Oryza sativa spontanea*) from Hainan Island is the source for cytoplasmic male sterility for hybrid rice in China, where hybrid rice now covers 7 million hectares and accounts for one-quarter of the area devoted to the cereal (Swaminathan, 1984b).

In some cases it is difficult to pinpoint specific ways in which wild species have contributed to improved agricultural production, because interspecific crosses have been made in a routine fashion for so long. In the case of sugar cane, for example, wild species have been an integral part of breeding since the 1880s (Brandes and Sartoris, 1936; Grassel, 1965; Daniels et al., 1975). An 1880 epidemic of sereh disease, probably caused by a virus, in sugar-cane fields on Java triggered a search for resistant material, and resistance genes were located in a wild cane (*Saccharum spontaneum*) and sugar-cane varieties outside of Indonesia. The inclusion of germplasm from *S. spontaneum* and exotic cultivars in sugar-cane breeding programs on Java led to the development of the famous, high-yielding noble canes (Brandes and Sartoris, 1936). Nearly all the major sugar-cane varieties grown throughout the world contain germplasm from *Saccharum spontaneum*, which has contributed resist-ance to several pathogens, including mosaic disease (Ramawas and Na-gai, 1984; Stalker, 1980). Virtually all of the commercial cultivars de-veloped by the Hawaiian Sugar Planters' Assocation (HSPA) since the early 1960s contain genes from wild canes such as *S. robustum*, *S. spon-taneum*, and *S. sinense* (Heinz, 1967). Wild germplasm in commercial sugar cane in Hawaii is a major reason for the stability and high pro-ductivity of cultivars in Hawaii; wild species have provided vigor and re-sistance to pests and diseases. These species have also contributed sig-nificantly to the productivity of sugar cane in the Caribbean.

Because of the manner in which the experiment station of the Ha-waiian Sugar Planters' Association (HSPA) does its crossing, it is diffi-cult to identify which wild species have contributed particular charac-teristics in Hawaii. The HSPA uses a polycross method to increase the number of crosses it can perform in a year. With this method, several different clones that are nearing maturation are cut and placed in a weak solution containing sulfuric acid, where they survive for about a month. The female parent then receives pollen from nearby clones. Seeds from the female parent are collected and germinated for screen-ing; over a million seedlings are tested each year, and only a tiny frac-tion eventually become commercial cultivars.

OVERCOMING BARRIERS TO WIDECROSSING

Wild species that are closely related to crops are relatively easy to cross, but others still challenge the skills and techniques of scientists. As a gen-eral rule, widecrossing is easier between species of the same genus; however, a few crop hybrids have been achieved between plants of dif-ferent genera. Thus far, though, crosses between species of different families have not been possible. Many of the routine widecrosses per-formed today by breeders, such as the maize x *Tripsacum* crosses at the International Maize and Wheat Improvement Center (CIMMYT— Centro Internacional de Mejoramiento de Maiz y Trigo), were not pos-sible until this century, when genetics and cytology have become better understood and microtechniques for manipulating plant germplasm were perfected.

One of the more common problems associated with widecrossing is the crossing of plants with unmatched chromosome numbers (different ploidy levels and different base chromosome numbers). It is easier to cross diploid plants that each contain two sets of chromosomes (2n) than, for example, to attempt to form a cross between a diploid (2n) and a tetraploid (4n), or where the parents of the widecross have different basic chromosome numbers, as in maize x *Tripsacum* crosses (maize has 20 chromosomes, whereas *Tripsacum* has 36). In the case of potatoes, the haploid (n) chromosome number (x) is 12; diploid potatoes thus have 24 chromosomes, triploids 36, tetraploids 48, pentaploids 60, and hexaploids 72 chromosomes (6x = 72). The common potato (*Solanum tuberosum*) is a tetraploid, but many useful genes are found in wild potatoes, three-quarters of which are diploids (N. Smith, 1983a).

Ploidy barriers can be surmounted in several ways. One way of crossing diploid and tetraploid potatoes is to obtain specimens of a diploid potato with unreduced gametes. In such plants normal reduction division in the formation of gametes (sexual cells) does not occur, and gametes are 2n instead of n. This results in egg cells with a full set of chromosomes (24) instead of half (12), as is normal. The unreduced gamete route has enabled breeders to introduce to the common potato genes coding for resistance to virus diseases and cyst nematodes from *Solanum chacoense* and *S. vernei*.

Another way to cross species with different ploidy levels is to double the number of chromosomes with colchicine, a chemical derived from autumn crocus bulbs. Diploid potatoes can hybridize with triploid and tetraploid species in this manner. Useful characteristics of wild, perennial species of *Medicago*, important forage legumes, have been transferred to cultivated species after artificial chromosome doubling (Von Borstel and Lesins, 1977).

A third manner in which breeders pass genes between plants of different ploidy levels is by employing bridge species. A bridge plant is able to cross with a wild species as well as the cultigen and is used whenever the latter two would not otherwise be able to exchange genes. A triploid bridge has allowed the passage of desirable genes from wild einkorn (*Triticum boeoticum*), a diploid grass, to durum wheat (*Triticum turgidum*), a tetraploid (Watson, 1970). Durum wheat has in turn served as a bridge to convey resistance to eyespot disease caused by *Pseudocercosporella herpotrichoides* to bread wheat (Doussinault et al., 1983). Severe eyespot infections lead to lodging (stem collapse), and the disease has caused severe yield reductions in wheat lands of the Americas, Europe, New Zealand, Australia, and Africa. Screened varieties displayed poor resistance to the fungal disease. High levels of resistance to eyespot disease were located in a wild grass (*Aegilops ventricosa*) that was introduced to hexaploid bread wheat via the tetraploid durum wheat. And a wild grass (*Aegilops speltoides*) was used as a bridge to transfer resistance to yellow rust (*Puccinia striiformis*) from a related wild grass (*A. comosa*), indigenous to the Aegean Sea area, to bread wheat (Riley et al., 1968).

Irradiation has also been used to facilitate widecrossing in certain cases (Harlan, 1976). High doses of short-wave radiation induce chromosome breakage, thereby permitting new recombinations. Cells of tomato about to divide (pre-meiosis) have been bombarded with X-rays to break down crossing barriers between two potato species (Rick, 1967). Progeny of a cross between *Triticum turgidum dicoccoides*, used as a bridge, and *Aegilops umbellata*, a source of resistance to leaf rust, have been irradiated to translocate the resistance gene from *A. umbellata* to a wheat chromosome (see Prescott-Allen and Prescott-Allen, 1983). Irradiation, though, is not commonly used in breeding for most crops, although it was thought to have more promise several years ago. A major problem with irradiation is that it can cause deleterious mutations and abnormalities. Artificially mutated strains can be helpful in breeding, but they can complicate the already difficult task of widecrossing.

Whichever method is used to overcome widecrossing barriers, considerable backcrossing to the crop parent is normally required to achieve an agronomically acceptable line. To improve the vigor of widecrosses, scientists at CIMMYT use different wheat parents in backcrosses. In some cases, initial crosses have to be repeated numerous times before fertile progeny are obtained. Seeds from the first cross between different species must sometimes be treated with a chemical called gibberellin to coax them to germinate (Dionne, 1963). Gibberellin is also used at CIMMYT to promote pollen tube growth and improve endosperm development in the F1 seed of wheat widecrosses. Furthermore, embryos from initial crosses may require careful *in vitro* culture to ensure their survival. For example, most hybrids between the cultivated tomato and a wild species, *Lycopersicon peruvianum*, a particularly useful source for improved vitamin C content and for resistance to spotted wilt, curly top, verticillium wilt, and root knot nematode, have had to be secured by embryo culture (Rick and Smith, 1953).

WILD SPECIES IN GENE BANKS

In spite of the importance of wild species in crop breeding, they have been generally neglected in germplasm collecting. Reasons for ignoring wild species during collecting trips include the fact that widecrossing is time consuming and often difficult, little or no information is available about potentially useful qualities such as insect or disease resistance, and crop breeders have been generally preoccupied with high-yielding material and earliness. Also, many wild grasses and legumes are particularly difficult to maintain in gene banks because seed yield is often low

TABLE 8.2
Wild species as a percent of total holdings in significant germplasm collections.

Crop	Held by International Agricultural Research Centers	Held in All Collections
CEREALS		
Rice	2.0	2.0
Sorghum	1.4	0.5
Pearl millet	0.1	0.2
Barley	0.001	0.5
Wheat	0	10.0
Maize	1.0	0.01
Minor millets	0.1	0.1
ROOT CROPS		
Potato	2.0	20.0
Cassava	1.0	1.2
Sweet potato	0	0.1
LEGUMES		
Beans	0.5	0.5
Chickpea		0.1
Cowpea	0	0
Groundnut	0.2	0.3
Pigeonpea	0.4	0.05

Source: IBPGR, 1983c:8.

during regeneration; seed cases shatter, thereby spilling their contents on the ground; and many wild species are photoperiod sensitive (Chang, 1976b:26; Harlan, 1976). Furthermore, wild species often mature over a prolonged period, which means that seeds have to be harvested on virtually a daily basis, which increases costs, or their flowering time does not synchronize with targeted cultivars.

Wild species typically account for less than 2 percent of gene bank accessions (Table 8.2). Only the wild relatives of wheat, potato (Figure 8.2), tomato, and, to a limited extent, maize have been extensively collected and preserved in gene banks, and work still remains to be done in obtaining more wild materials for crossing programs for those important crops. The International Board for Plant Genetic Resources (IBPGR) is giving special emphasis to the collection of wild species in future missions (IBPGR, 1983b:7).

Wild species are making an ever more important contribution to crop improvement (Harlan, 1976; Knott and Dvorak, 1976). For some crops, such as maize, wheat, and barley, wild species are only occasionally employed in breeding programs. But with other crops, particularly the common potato, forage and forestry species, wild germplasm is frequently tapped. As more wild species are collected and evaluated, wild relatives of crops will be increasingly used in breeding programs. Phillips (1976), for example, argues that the trend in cotton breeding is toward greater exploitation of wild species, partly in response to a desire to avoid environmental damage resulting from heavy pesticide applications on many cotton farms. Breeders generally see an increased use of wild species for resistance to diseases. For instance, now that wild germplasm of groundnut has been better collected and evaluated, and scientists have managed to cross groundnut with thirteen wild species, breeders for that crop are likely to turn increasingly to wild species for pest and disease resistance (ICRISAT, 1985). Wild relatives of groundnut often contain the only genes for resistance to certain diseases (Moss, 1980).

Developments in biotechnology will also lead to a greater use of wild species in breeding work. Many species that cannot currently be crossed, even with techniques such as chromosome doubling and the use of bridge species, are likely to be able to exchange genes when recombinant DNA methods are more advanced (Wolf, 1985). When protoplast fusion is more precise and effective vectors for recombining DNA are identified, for example, genes may be passed between biologically widely separated plants.

Biotechnology will not, however, make germplasm collections of wild species and crop varieties redundant (Peacock, 1984). Nor is it likely that the time needed to develop suitable lines from widecrosses will be significantly shortened. It will still take patience and numerous backcrosses to develop desirable varieties. And although it is now possible to synthesize some genes in the laboratory, germplasm collections are still essential.

Scientists need models of naturally occurring genes in order to decode them for copying. Thus biotechnology developments will enhance the value of gene banks. Furthermore, some plants are on the verge of domestication, and more could eventually be cultivated (Hinman, 1984). Of the 200,000 species of flowering plants, 5,000 are used for various purposes, but only about 500 species have been domesticated and fewer than 150 are significant in commerce or subsistence (Wilkes, 1984).

It is essential that as much wild germplasm as possible be collected before it becomes extinct. While it is true that some DNA has been re-

8.2. Wild species of tuber-bearing *Solanum* potatoes being harvested in evaluation trials at the International Potato Center, Lima, Peru, 1983.

covered from the hide of an extinct mammal, only a tiny fraction of its genetic code was intact. It seems unlikely that scientists will be able to recreate living organisms from preserved specimens, whether pickled in a bottle or pressed between herbarium sheets. As the title of the book edited by Prance and Elias (1977) suggests, extinction is forever.

A CASE STUDY IN RICE GERMPLASM: IR36

Because not many people have the opportunity to learn just how genetic resources are used to improve our crops, we will tell, in detail, the story of how cultivated and wild rice stocks were used to produce the most widely planted rice variety in history, IR36. It is a fascinating account, but it is not unique; similar histories could be written for popular varieties of wheat, maize, or grain legumes. But the IR36 story is a good example of how plant breeders and other agricultural scientists work to produce the food crops that sustain us. The crop became the world's most widely cultivated rice within just a few years, and its rapid adoption is especially important because of the large number of people it nourishes.

The IR36 story is also significant because the semi-dwarf variety is the product of international cooperation. IR36 was produced by the International Rice Research Institute (IRRI) in collaboration with developing countries and industrial nations. Rice lines from six countries are in the IR36's genealogy (Figure 9.1). This variety provides a good example of the trend in agricultural research towards a global enterprise and illustrates the importance of gene banks and international cooperation in germplasm conservation and breeding. The IR36 story also underscores the value of screening germplasm in "hot spot" locations where disease and insect pressure are particularly intense, and illustrates the use of elite and wild material in breeding.

BREEDING MODERN RICE

The modern era of rice breeding has progressed rapidly since World War II, and particularly since IRRI's establishment in 1960. Rice breeding and research were greatly stimulated by IRRI's success in developing IR5 and IR8, short-statured rices that were high yielding and fertilizer responsive without keeling over because of weak stems. IR8 dominated rice production in tropical Asia during the 1960s (Swaminathan, 1982), but after its release it soon became obvious that more disease and insect resistance was needed. Sources of resistance to major pests and diseases were sought in IRRI's rice germplasm collection.

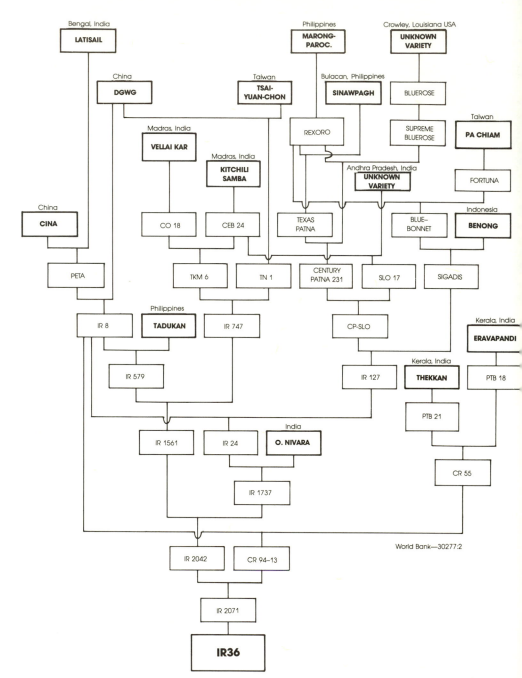

9.1. Ancestry of IR36. (Information from
W. R. Coffman, pers. comm.)

This collection was still being assembled in the 1960s, growing from an initial 266 lines that were first planted at IRRI during the 1961-62 dry season to over twelve thousand accessions in 1970 (Chang, 1980). Careful screening identified lines with resistance to such production constraints as green leafhopper (*Nephotettix virescens*), tungro virus, bacterial blight, and blast. The aim was to combine as much resistance as possible to such pests while retaining the high-yielding, fertilizer-responsive, and short-statured features of IR8. A series of new varieties with greatly increased resistance to pests and diseases was thus developed over the next two decades (Table 9.1).

During 1968 and 1969, IR8 fields suffered serious bacterial blight attacks, and in 1970 tungro virus caused widespread losses across the Philippines. Furthermore, people in some areas complained that IR8's grain quality was inferior. IRRI's breeding program, in close cooperation with national programs, addressed these concerns. Each of the varieties developed after IR8 was launched in 1966 had more suitable attributes, at first in improved grain quality, and then in greater resistance to certain pests and diseases.

IR20, released in 1969, replaced IR8 within a couple of years, but was itself severely damaged by biotype 1 of the brown planthopper (*Nilaparvata lugens*) and grassy stunt virus in 1973. IR26, which was resistant to brown planthopper, was released in the same year, and by 1975 it had risen swiftly to dominate rice production in the Philippines. It also became the most widely planted rice variety in Indonesia and Vietnam. But again, the ability of pests and diseases to catch up with crop varietal advances became evident when a new brown planthopper biotype appeared that could not be controlled by the type of resistance built into IR26. Fortunately, scientists had anticipated the evolution of another brown planthopper race, and in 1976 IR36, with its wider resistance to brown planthopper, was ready for release. IR36 replaced IR26 as the leading variety within a year and now is the mostly widely sown rice in the Philippines (Figure 9.2). By 1982, IR36 covered 11 million hectares of Asia's rice lands, making it the world's most widely planted rice variety in history (Swaminathan, 1982).

THE JOURNEY TO IR36 AND BEYOND

Keeping ahead of existing and new pests, diseases, and crop management difficulties requires an energetic and well-supported research team with expertise in such areas as plant pathology, entomology, plant physiology, agronomy, soils, and plant breeding. Interdisciplinary teams must be able to identify key existing problems and find ways to

TABLE 9.1

Rice varieties developed by IRRI in collaboration with national programs, 1966-1985, with reactions to certain diseases and pests.

Variety	Year Released	BPH Biotypes			Grassy Stunt	Tungro	GLH	Blast	BB	SB
		1	2	3						
IR8	1966	S	S	S	S	S	MR	S	S	S
IR20	1969	S	S	S	S	MR	R	MR	R	MR
IR22	1969	S	S	S	S	S	S	S	R	S
IR24	1971	S	S	S	S	S	R	S	S	S
IR26	1973	R	S	R	S	MR	R	MR	R	MR
IR28	1975	R	S	R	R	R	R	R	R	MR
IR29	1975	R	S	R	R	R	R	R	R	MR
IR30	1975	R	S	R	R	R	R	MS	R	MR
IR32	1975	R	S	R	R	R	R	MR	R	MR
IR34	1975	R	S	R	R	R	R	R	R	MR
IR36	1976	R	R	S	R	MR	R	R	R	MR
IR38	1976	R	R	S	R	R	R	R	R	MR
IR40	1977	R	R	S	R	R	R	R	R	MR
IR42	1977	R	R	S	R	R	R	R	R	MR
IR44	1978	R	R	S	S	R	R	MR	R	MR
IR46	1978	R	S	R	S	R	R	R	R	MR
RI48	1979	R	R	S	R	R	R	MR	R	MR
IR50	1979	R	R	S	R	R	R	MS	R	MR
IR52	1980	R	R	S	R	R	R	MR	R	MR
IR54	1980	R	R	S	R	R	R	R	R	MR
IR56	1982	R	R	R	R	R	R	R	R	MR
IR58	1983	R	R	S	R	R	R	R	R	MR
IR60	1983	R	R	R	R	R	R	R	R	MR
IR62	1984	R	R	R	R	MR	R	MR	R	MS
IR64	1985	R	MR	R	R	R	R	MR	R	MR
IR65	1985	R	R	S	R	R	R	MR	R	MR

Abbreviations: BPH = Brown planthopper; GLH = Green leafhopper; BB = Bacterial blight; SB = Stemborer; S = Susceptible; MR = Moderately resistant; R = Resistant

overcome them; in addition, they must be able to anticipate future problems and work out methods to handle them, preferably before agricultural productivity dips. Good research on the biology and potential spread of pests and diseases is thus essential, as is quality work on management problems that commonly accompany the adoption of modern, higher-yielding technologies. It is almost axiomatic that the higher yields become, the greater the effort needed to sustain the gains (Pluck-

9.2. Harvesting IR36 near Los Baños, Philippines,
January 1985.

nett and Smith, 1986b). Low-yielding agriculture may be more stable, but it is often so unproductive that it is unprofitable and incapable of feeding rapidly growing urban and rural populations.

During the 1960s, scientists identified blast disease, green leafhopper, bacterial blight, and tungro virus as important constraints to rice production and accordingly studied their biology and virulence patterns intensively. In addition, after screening germplasm they located some resistance to these problems. They then passed resistant materials along to plant breeders so that the desirable characteristics could be incorporated into new, more robust, high-yielding lines. Whenever possible, they sought more than one genetic source of resistance to each pest or disease to increase the chances of longer-lasting resistance. Pathologists and entomologists also checked minor diseases and pests in fields that might become serious problems in the future. Brown planthopper and grassy stunt were singled out for special attention because rice farmers in many areas were able to plant several crops each year after the introduction of IR varieties with their ever-shorter growing periods; accelerated cropping cycles within the same variety create a propitious environment for the rapid evolution of pests and diseases.

The brown planthopper not only causes "hopper burn" as a result of its rapid suction of plant sap, but it also transmits grassy stunt virus. The relatively minor viral disease was therefore also selected for intensive study.

As it turned out, the choice of grassy stunt virus for close scrutiny was a wise one since the pathogen later caused widespread disease outbreaks in Asian rice fields. When young rice plants are infected with grassy stunt at an early stage, they generally do not flower and thus produce no grain; the virus can cause acute or even total yield loss in infected fields (Khush and Ling, 1974). Susceptible plants tiller excessively, have narrow, yellowish green leaves with rusty spots, and are severely stunted. Brushlike diseased plants produce either no seed heads or small panicles with deformed grains (Khush et al., 1977).

Grassy stunt disease, transmitted by brown planthopper, became a serious problem in parts of the Philippines, Thailand, Sri Lanka, India, Malaysia, and Indonesia in the late 1960s and 1970s (Khush *et al.*, 1977). Rice fields in Vietnam, Kampuchea, Laos, and Burma were probably also affected by the virulent disease (Figure 9.3). In Indonesia, the first recorded outbreak of grassy stunt virus disease occurred in Central Java in 1969, and by the early 1970s the epidemic had spread across more than 116,000 hectares of rice fields. Between 1974 and 1977 losses to the disease and its vector in Indonesia exceeded 3 million tons of paddy rice, worth more than $500 million (Palmer et al., 1978; Hibino, 1984). In 1977 alone, rice production losses due to the disease and insect pest reached 2 million tons, enough to feed 6 million people for a year (Conway and McCauley, 1983).

An extensive search for material resistant to grassy stunt was conducted at IRRI's gene bank in 1970. After screening five thousand accessions and one thousand breeding lines, only one accession of a wild rice, *Oryza nivara*, collected in Uttar Pradesh, India, in 1963 by a scientist of the Central Rice Research Institute at Cuttack was found to resist the disease (Chang, 1976c; Khush et al., 1977) (Figure 9.4). Within the single accession resistant to grassy stunt, only three plants contained a gene for resistance to the disease (Chang, 1985). The accession of resistant *O. nivara* made its way to IRRI as part of a rice collection provided by the Central Rice Research Institute.

The resistant wild rice grows in natural and disturbed habitats in India and neighboring parts of Asia. However, *O. nivara* has more undesirable traits than favorable ones. Seed shattering at maturity, droopy leaves, a spreading growth habit, long awns (stiff bristles protruding from the seed case), red pericarp (seed coat), and numerous sterile seeds are among its unwanted characteristics. Besides its resistance to

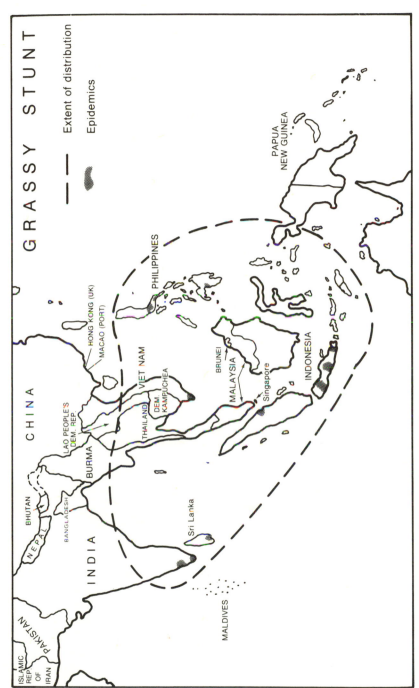

9.3. Approximate distribution of grassy stunt virus disease and areas where epidemics have occurred in rice fields. (From Khush et al., 1977.)

177

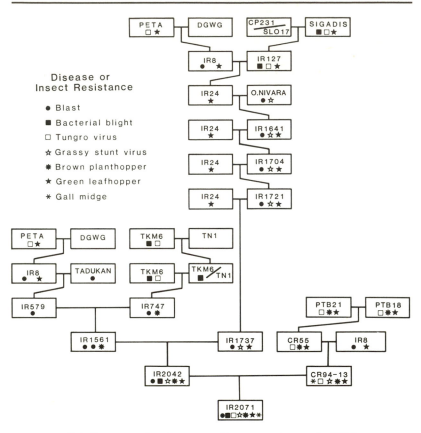

9.4. Sources of resistance to diseases and pests in IR36
(IR2071). (From Swaminathan, 1984b.)

grassy stunt virus, *O. nivara*'s desirable traits include high tillering and resistance to blast disease (Khush et al., 1977).

During screening and breeding work, IRRI found that a single dominant gene conferred resistance to grassy stunt. The gene was designated Gs, and it was found to segregate independently of the gene for resistance to brown planthopper, the vector of grassy stunt. The dominant gene for resistance to brown planthopper was designated as Bph1 (Khush et al., 1977).

A means of rearing the insect in the laboratory had to be developed in order to elucidate the life cycle and other attributes of brown planthopper and grassy stunt virus. Its breeding dynamics and feeding habits were studied in detail. Experimental fields were artificially infested with the minute brown insects, and the virus transmission ability of the planthopper vector was determined. Screening of the rice germplasm collection for resistance to brown planthopper and grassy stunt virus began. As is usual in such work, screening methods had to be developed to ensure reliable selection of lines with varying degrees of resistance to the planthopper or virus. Special techniques were required for virus-transmission studies.

Brown planthopper resistance was found in two lines (PTB18 and PTB21) from southern India. A major step was achieved towards in-corporating brown planthopper resistance into other varieties when scientists in India crossed the two lines to produce CR55. Another source of resistance to the insect came from the cross of TKM6 from Madras, India, and Taichung Native 1, a variety from Taiwan, that produced IR747.

THE BREEDING OF IR36

Thirteen rice varieties from six countries and a wild species related to cultivated rice were utilized in the process (Figure 9.1). In making the crosses, IRRI plant breeders, led by Gurdev S. Khush (Figure 9.5), used the basic information about germplasm resources, screening techniques, and the biology of disease and pest organisms gained by other scientists. The initial crosses that led to the development of IR36 were made in 1969. By late 1970, IR1561 had reached the fifth-generation stage and showed resistance to bacterial blight, stemborers, and brown planthopper. However, it was still susceptible to green leafhopper, grassy stunt, and tungro virus.

In July 1969, *O. nivara* was crossed with IR24 to incorporate the gene for grassy stunt resistance into cultivated rice. The first generation (F1) plants were inoculated artificially at the seedling stage, and all were found resistant. These were backcrossed to IR24, and the first back-cross progenies were again inoculated with grassy stunt at the seedling stage. Resistant plants with good agronomic characteristics, such as minimal shattering of seed, low sterility, and dwarfness, were used for making the second backcross with IR24. Most of the plants from this backcross were dwarf, had many shoots (high tillering), had no bristles protruding from the seed cases (awnless), did not shatter, and were highly fertile. The third backcross progenies, designated IR1737, were identical to the recurrent parent IR24 (Khush et al., 1977) and resisted grassy stunt, green leafhopper, and blast. In addition, they had good grain quality and excellent yield potential. Still, they were susceptible to brown planthopper, bacterial blight, and tungro virus.

In early 1971, a selection of IR1737 was crossed with a selection of

179

9.5. Gurdev S. Khush, senior rice breeder at the International Rice Research Institute, Los Baños, Philippines, and a leading figure in the development of IR36, the world's most widely planted rice variety.

IR1561 to create IR2042. The latter was in turn crossed with a gall midge and tungro-resistant line from India (CR94-13) to produce IR2071. IR36 was selected from this cross. The complex evaluation procedures that led to the selection of IR36 from this cross are outlined below.

The F1 seeds of IR2071 were planted in January 1972, and the seedlings were again inoculated with grassy stunt. The seeds were harvested from the resistant seedlings in April 1972. The F2 population of IR2071 was sown in mid-1972 without insecticide protection at the Maligaya Rice Research and Training Center, Nueva Ecija, Central Luzon, Philippines. Tungro disease was widespread in Central Luzon in 1972. Stemborer populations were also higher there than at IRRI. Plants showing tungro or stemborer damage were culled. Resistant plants harvested to make up the F3 generation were planted out at IRRI and exposed to heavy pressure from brown planthopper, bacterial blight, blast, and green leafhopper, using techniques designed for screenhouse and field conditions.

A special screening technique for grassy stunt was employed in the F4 generation (Figure 9.6). IR24, highly susceptible to both brown planthopper and grassy stunt, was used to build up large brown planthopper populations and to serve as a check variety. Field evaluations were also made for bacterial blight, blast, and leafhopper. An early maturing line, designated IR2071-625-1, appeared vigorous and showed no symptoms of grassy stunt. A later assessment showed it also resisted bacterial blight, blast, green leafhopper, and brown planthopper. Seeds of IR2071-625-1 were bulk harvested in September 1973.

A small seed-increase plot was planted in September 1973 and harvested in February 1974, when four hundred plant selections were made. These selections were planted in individual plots in late February 1974. Plot 252 appeared uniform and at harvest in May 1974 was designated IR2071-625-1-252. This reselection from IR2071-625-1 was entered in IRRI's replicated field trial in June 1974 and was tested in other coordinated trials.

In the meantime, the incidence of tungro in the Philippines had receded considerably by the end of 1973, making it impossible to screen for the disease in that country. Seeds of F5 lines of IR2071, together with several thousand other breeding lines, were therefore sent to Indonesia in January 1974 to be planted at Langrang in South Sulawesi, where tungro was prevalent. The resistance of IR2071-625-1 to tungro was confirmed in that trial.

In May 1974, seeds of promising F6 lines were sent to the Central Rice Research Institute in Cuttack, India, to screen for resistance to gall midge. Gall midge populations increase markedly during September and October, and IR2071-625-1-252 withstood heavy pressure during those months. Meanwhile, F6 lines were screened at IRRI for resistance to stemborers during the July-October 1974 growing season.

This strategy of screening in disease and insect "hot spots" with the cooperation of several national programs made it possible to verify the multiple resistance of IR2071-625-1-252 to blast, bacterial blight, grassy stunt, tungro, green leafhopper, brown planthopper, stemborer, and gall midge by the end of 1974. In the same year, two growing seasons of testing in replicated yield trials had established the high yield potential of the line. The line also had good grain type and a high milling recovery.

By early 1975, IR2071-625-1-252 was ready to be entered in the Philippine Seed Board Lowland Cooperative Performance Tests. During two growing seasons in 1975, the promising line outyielded all other early entries in these trials. In March 1976, the Rice Varietal Improvement Group of the Philippine Seed Board recommended the naming of IR2071-625-1-252 as IR36. This recommendation was approved by the Seed Board in May 1976, and IR36 was ready for widespread planting.

Within a year of its release, IR36 replaced IR26, then the dominant rice variety in the Philippines. By 1980-81, 2.73 million hectares, 78

9.6. Planting trials to evaluate rice lines. The International Rice Research Institute, Los Baños, Philippines, 1968.

percent of the rice-growing area of the Philippines, was sown to modern, high-yielding varieties. Of the area planted to modern varieties, 90 percent was occupied by IR36.

IR36 was also welcomed by farmers in other countries. In Indonesia, for example, IR36 was recommended in 1977 and became the most widely planted rice variety in the country within a year. By 1980-81, IR36 occupied 5.3 million hectares of the 9.3 million hectares devoted to rice in the country. In Vietnam, brown planthopper has been a serious pest, particularly when biotype 2 appeared in 1977. In that year,

Vietnam embarked on a crash program of furnishing IR36 to farmers after importing more than 250 tons of IR36 seed from the Philippines. By 1981, IR36 covered some 2.1 million hectares. Most of the IR36 was in South Vietnam, but some was also planted in North Vietnam.

In India, IR36 was recommended for Orissa in 1977 and West Bengal in 1978. IR36 now grows in West Bengal, Kerala, Andhra Pradesh, Orissa, and Madhya Pradesh, and was recommended as an all-India variety by the Central Variety Release Committee in 1981. Significant areas are also sown to IR36 in Kampuchea, Laos, Bangladesh, Sri

183

Lanka, and Malaysia. IR36 is also a recommended variety in several African countries, such as the Central African Republic, Malagasy Republic, and Zambia.

ASSESSMENT

Several factors account for IR36's spectacular take-off. First, IR36 is early-maturing, with seeds ready for harvest after 107 to 110 days, and high-yielding. Second, IR36 grains are long, slender, and translucent with good milling qualities. Third, it resists many of the major rice pests and diseases, including green leafhopper, brown planthopper, stemborer, blast, bacterial blight, tungro, and grassy stunt. Fourth, IR36 resists gall midge in India and Sri Lanka. Further, 1R36 tolerates soil salinity, alkalinity, iron and boron toxicity, and zinc deficiency in wetlands and tolerates iron deficiency and aluminum toxicity in drier regions. IR36 also survives moderate drought.

Thus far, IR36 yields are relatively stable. The variety consistently produces between 4 and 6 tons per hectare in farmers' fields and up to 9 tons per hectare in experimental trials. Indeed, it is estimated that IR36 is responsible for an additional 5 million tons of rice each year (Kahn, 1985:242). However, IR36 cannot be relied upon indefinitely. Even superstars in the varietal relay race eventually succumb to new diseases, pests, or other environmental challenges. Already in some parts of India, the Philippines, Thailand, and Taiwan, varieties with the resistance gene from *Oryza nivara* are showing signs of infection with grassy stunt, suggesting that new strains of the virus have evolved (Hibino, 1984). A new virus disease, ragged stunt, and another unknown virus started attacking IR36 in certain areas in 1982 (Lewin, 1982). Newspaper reports indicate that IR36 is becoming susceptible to tungro virus in certain parts of the Philippines.[1]

After a decade of occupying vast areas of Asia, IR36 can be expected to be threatened by mutations of existing pests and diseases or the appearance of new ones. Large areas occupied by a few varieties do not necessarily lead to yield instability—although the danger is always there—provided that replacement lines are ready for release when current cultivars are no longer holding up. Ironically, the very success of IR36 will contribute to its downfall.

The downturn of IR36 is on the horizon, but recently-developed high-yielding varieties, such as IR60, IR62, and IR64, are ready to replace it (Table 9.1). Successors to IR36 are resistant to an even broader

[1] *Times Journal*, Manila, Philippines, 21 September 1984.

range of diseases and pests, and they mature faster. Within a decade, IR36 may well be gone, but it has contributed enormously to the feeding of Asia's population. It will stand as a landmark variety attesting to the skill of scientists using genetic resources in exciting and innovative plant breeding, and its genes will form the foundation of future named releases.

GLOBAL IMPERATIVES

Gene banks are largely a twentieth-century effort. Before 1900, most germplasm resources were held by farmers, and governments and scientists played a minor role in preserving and exploiting plant genetic diversity. For a few crops, notably tropical cash crops, botanic gardens, breeding programs operated by colonial powers, and private collectors held the major germplasm collections.

In the early 1900s, the Soviet Union and the United States initiated major plant-breeding programs and began to move towards more systematic collecting and use of crop germplasm. Early plant gathering expeditions were often extensive, as in the case of Vavilov's numerous trips, but the germplasm was held essentially as working collections for the benefit of breeders. In response to the incipient loss of traditional varieties and wild relatives in their natural areas, some germplasm collections were placed in cool storage beginning in the early 1940s.

The impressive growth of gene banks, particularly within the last decade, was stimulated in part by a series of expert meetings organized by the Food and Agriculture Organization (FAO) of the United Nations in the 1960s. These meetings highlighted the importance of conserving genetic resources during a period of rapid and profound ecological and cultural changes in places where many crops originated. After scientific and public attention to the problem was aroused, the needed international instrument for following through on a program of action emerged with the founding of the International Board for Plant Genetic Resources (IBPGR) in 1974.

The development of a global network of gene banks has been the primary mission of IBPGR since its inception. This network of gene banks includes long-term collections under reduced temperatures as well as field gene banks for plants with seeds that do not tolerate drying and freezing or for crops that are propagated vegetatively. Only five or six long-term gene banks were functioning in 1974, and most of those were in industrial countries. In 1986, some fifty such gene banks were operational, with more under construction or in the planning stage. During its first ten years of existence, IBPGR has sponsored the collecting and storage of over 100,000 germplasm samples.

Other gains have accompanied the increased efforts to collect and safely store germplasm. A significant part of the world scientific community has been mobilized in advising, planning, and conducting plant genetic resources activities. Hundreds of individuals have been trained to carry on the necessary work. Germplasm moves regularly between virtually all countries, and today most plant breeders have much better collections of materials to draw on.

Despite the impressive strides in collecting, much remains to be done. Early collection efforts tended to follow an opportunistic pattern, garnering a varied but often incomplete sample of the available genetic diversity. Today, collecting tends to be targeted to specific areas in order to fill gaps, to gather material threatened with extinction, or to obtain plants with particular traits. Although germplasm collection of some of the major cereal crops is nearly complete, much remains to be collected for other crops.

Gene banking has come a long way, but problems remain there as well. Many collected materials have not been properly characterized or evaluated. Documentation of collections is often incomplete. Gene banks without properly characterized and evaluated holdings are of limited use to breeders and a drain on scarce funds for agricultural research. Not all gene banks work well and long-term financing for most germplasm collections is uncertain. Some gene banks do not function adequately because of high electricity costs or malfunctioning equipment. Too many gene banks are more like stuffy hothouses than cold-storage facilities for seeds.

A widely shared concern among germplasm specialists is the inadequate knowledge base for genetic resources work. Little or no research is being conducted on some existing and emerging problems with gene banks, such as stability of seeds under reduced humidity and temperature, seed health problems before and during storage, handling of seeds that do not tolerate freezing, and the size and range of samples to be stored.

Fortunately, scientists, donors, and governments are increasingly aware of the need to conserve and make better use of crop germplasm. The overall picture is now much brighter than in the early 1970s, when concern over the narrowing genetic base of our crops spread among agricultural scientists and alarmed the general public and politicians. It is now clear that the effort to gather the germplasm of crops and wild relatives must continue to increase the options for improving agriculture.

GENE BANKS AND DEVELOPING COUNTRIES

Much of the germplasm of crops and wild relatives that remains to be collected is in the tropics. Because Third World leaders are besieged by many pressing problems, they often do not regard the setting aside of scarce financial resources to gather, store, and evaluate plant germplasm as a top priority. Still, new technologies are emerging that make it feasible to store an ever wider range of crops in *ex situ* gene banks with reduced temperatures, and developing countries must be urged to make use of them. These gene banks can be practical and prudent investments, and medium-term facilities are affordable and desirable for most nations. All countries have a stake in preserving plant germplasm and must be encouraged to take an active part in the important task of gathering and utilizing plant genetic resources.

Gene banks can be adapted for the many small nations with limited funds and small germplasm reserves. Well-maintained freezer chests and seed dryers suffice for modest collections. Most small nations do not have enough plant-breeding expertise to use germplasm effectively; a top priority for such countries should be to develop a better plant-breeding capability. Unless the germplasm endowment is large, or the country has a particular desire to preserve specific germplasm for its needs in long-term storage, more affordable medium-term facilities should meet the needs of most breeders.

Indeed, many developing countries have already built, or plan to build, such modern gene banks. Some nations, such as Indonesia, have established national plant genetic resource committees and have drawn up long-term work plans. Mexico is expanding its genetic resource conservation efforts; it has built a long-term germplasm storage facility at Zacatecas, is putting up a medium-term gene bank at Chihuahua, and plans to erect another such facility at Oaxaca or Veracruz. Brazilians are also investing heavily in their gene bank at Brasilia, and their national genetic resources agency, CENARGEN (Centro Nacional de Recursos Genéticos), is an outstanding example of a well-equipped and staffed research and conservation facility.

The increase in gene banks in developing countries is significant. IBPGR has sought to redress the former imbalance in germplasm collections when most collections were held by industrial countries. Now the "South" has become well endowed with gene banks as well (Ford-Lloyd and Jackson, 1984).

SEEDS OF DISCORD

A debate has been stirring recently over who owns, controls, and benefits from germplasm collections. Much of the discussion centers on the acquisition and release of gene-bank accessions (Mackenzie, 1983; Mooney, 1983; Tucker, 1984; Walsh, 1984). Some complain that gene banks are mostly under the thumb of industrial nations and are beholden to private companies. Critics of the current system of gene banks often ignore the substantial flow of plant genetic material from the industrial countries to the Third World as well as their financial support to the international agricultural research centers that provide the bulk of the food crop germplasm to developing countries.

To counteract the perceived bias in the current system of gene banks, some individuals have suggested that IBPGR must be controlled by some "neutral" organization, such as the United Nations. Donors (some of them Third World governments) to IBPGR and sister international agricultural research centers within the Consultative Group on International Agricultural Research (CGIAR), itself a "neutral" organization, are concerned that such a move might compromise IBPGR's independence and autonomy.

In 1981, certain Third World governments, led by Libya, Mexico, and Peru, proposed that the current network of gene banks be brought under the jurisdiction of FAO, and this idea was discussed at the FAO biennial conference in November 1983 (FAO, 1983a,b). Member countries of FAO agreed to launch a legally nonbinding International Undertaking on Plant Genetic Resources, and an FAO International Commission on Plant Genetic Resources was established to oversee the Undertaking. The International Commission met for the first time in Rome in March 1985. An original notion to create a single international gene bank to house all the world's crop germplasm was soon dropped because it was deemed unrealistic and scientifically unmanageable.

The International Undertaking established principles to be adhered to by countries in the collection, maintenance, and exchange of plant germplasm. The Undertaking includes elite and current breeders' lines, as well as special genetic stocks. This inclusion of improved and experimental material has aroused concern in countries that recognize plant variety rights. But IBPGR has always stressed that medium- and long-term gene banks should not clutter their collections with breeders' lines. Gene banks should concentrate on primitive varieties, wild relatives, and obsolete cultivars, thereby housing a wide spectrum of genetic variability.

The Undertaking suggests that present international arrangements be further strengthened and legalized to ensure that an internationally coordinated network of national, regional, and international centers develops, including an international network of base collections in gene banks, under the auspices and jurisdiction of FAO. It is not clear how the present network would be amalgamated under FAO control, so the International Commission has requested studies of the matter.

Thus far, the International Undertaking has made little progress.

Some key member countries of FAO, such as the United States, do not support the Undertaking, although they agree with some of its principles. Other countries, such as Brazil and India, that were originally receptive to the Undertaking, are now having second thoughts. Furthermore, the Soviet Union is not a member of FAO and thus will not necessarily abide by any FAO decisions. It is difficult to assess the potential impact of the Undertaking on those countries that do not abide by it, nor is it easy to gauge how it will affect current collaborative efforts. Bilateral agreements on plant germplasm, such as that between China and the United States, could be complementary instruments. Brazil appears to be moving in this direction, at least with regard to export crops such as cacao and cashew (*Anacardium occidentale*). Similarly, the German Democratic Republic has worked out bilateral agreements with Iraq, Italy, and Mongolia to exchange plant germplasm through relevant academies. Such agreements often lead to the exchange of personnel and to other mutual benefits.

Concern has arisen that dissatisfaction over the handling of crop genetic resources in the international arena will create uncertainty in the donor community, resulting in reduced support for conservation work (Wolf, 1985). Also, protracted discussion and debate within FAO might lead to paralysis precisely when more collecting, evaluation, and exchange of germplasm is most needed (Ford-Lloyd and Jackson, 1984).

Whereas some problems remain with the current gene-banking system, much goodwill has been nurtured by it. International collaboration in germplasm acquisition and exchange is a reality. We consider that efforts would be better spent on improving the current system with increased financial support and training opportunities.

However the global network of gene banks operates in the future, whether under a legal umbrella provided by FAO or informally as now, the IBPGR, or a similar organization, needs to set standards, develop the appropriate technologies through strategic research, and act as a nonpolitical body working with programs in all countries. To be effective the organization must be scientifically sound, have credibility throughout the world, and enjoy solid financial support.

FUTURE TASKS

Specific deficiencies in current efforts to conserve and utilize germplasm are outlined in earlier chapters. Our aim here is to highlight some of the broad areas requiring urgent attention by researchers.

Seed physiology, seed health, plant ecology, population genetics, taxonomy, cryopreservation, and tissue-culture preservation are especially critical areas that need help. The effects of long-term preservation on crops are not well understood, nor are the storage conditions necessary for some crops, particularly those with seeds that cannot be easily dried and frozen. In addition to different storage and handling requirements among different crops, varieties within a crop destined for germplasm storage sometimes require special attention. At the Institute of Plant Breeding at the University of the Philippines, Los Baños, for example, scientists have found that not all banana varieties can be cultured with standard *in vitro* techniques.

More research is needed on collection, characterization, evaluation, multiplication, storage, and enhancement of germplasm so that it will be used more effectively. Seed health problems persist; indeed, some gene banks may store crop pathogens more successfully than seeds. In addition, more needs to be learned about how much genetic variability for each crop needs to be stored; in other words, how wide should the net be cast? A better understanding of the variation present in field populations is the key to answering this question. IBPGR has always emphasized collecting, but many more studies of the ecogeography of crop plants and their wild relatives are also needed. In the past, collecting has been widespread and generalized to build up substantial germplasm collections; in the future, collecting missions will be more selective, concentrating on priority areas and wild gene pools related to crops.

Partly because of the debate over germplasm ownership and distribution, many nations have become increasingly aware of the need for crop genetic conservation. Besides taking into account the needs for forestry, domestic animals, and microbes what steps should nations take in the immediate future to assure such conservation?

First, although it is possible to conceive of national, regional, and international gene banks all wedded into an international network, experience has shown that the regional approach is often doomed to failure. The Nordic gene bank is a successful model, but in its case the five cooperating countries share many political and cultural attributes. Relations between countries in close proximity are actually often tense and unstable; witness the weakness or collapse of many regional airlines and economic associations. In few parts of the world is there a group of nations that would willingly pool resources for germplasm work. Today a pattern of national programs backstopped by international agricultural research centers appears more realistic and thus prevails.

Second, not all nations need to establish and maintain long-term gene banks. Many Third World countries have more urgent needs, and they would be better served by aggressive plant introduction programs and

191

by upgrading their crop-breeding capacity. After the attention given to gene banks during recent international meetings, some countries, which at this point hardly need collections maintained in perpetuity, have decided to develop long-term gene banks. For the foreseeable future, an ideal system would consist of strong international genetic resource centers linked with a number of well-run national programs.

Third, any country that has agreed to fund and maintain a gene bank must make a commitment to sustain the effort. Few countries have been willing to accept, or even consider, the minimal requirements already defined. Adherence to the FAO Undertaking does not ensure any ongoing commitment. Gene banks that are enthusiastically set up with donor support can easily come to an untimely end if the necessary long-term commitments do not materialize. This is one reason why IBPGR emphasizes the duplication of collections.

Fourth, all programs must provide more than lip service to the principle of free availability of materials to *bona fide* users. It remains to be seen whether the new FAO commission will identify defaulters and whether any action will be taken. For decades, a policy of goodwill has resulted in relatively free availability of germplasm; such informal cooperation is hard to legislate.

Fifth, a new attitude and fresh ideas are needed by gene-bank curators around the world. As a first step, curators must use passport data to sort out their accessions and to establish their origins. The U.N. Development Program (UNDP) and IBPGR have initiated a special cooperative project in Europe to further such work. It is better to have five hundred accessions gathered systematically from widely separated habitats than to have five thousand haphazardly collected accessions. Junk holdings complicate the work of curators, escalate conservation costs, and waste the time of breeders.

When the origins of accessions are ascertained, curators should see that collaborative arrangements move quickly to fill collecting gaps so that the net is cast wide. It is not helpful when curators emphasize the numbers of accessions under their management, as if numbers alone indicate the value of a collection; the same principle is true for libraries. Each gene-bank accession must be justified, just as library acquisitions must be selective, because of space, budget, and usage considerations.

Gene banks thus far have focused on crops, particularly food crops, but germplasm conservation also applies to livestock, forestry, medicinal plants, and microorganisms. Livestock gene banks, for example, are only now being given serious thought (Brown, 1984). Semen and embryos of some mammals can be frozen and stored for several months, or even years, but this development could be a double-edged sword.

Like the use of improved, widely adapted crop varieties, frozen embryo technology may lead to a narrowing genetic base of livestock herds. By administering hormones, desirable cows can be induced to produce an unusual number of eggs; these in turn can be fertilized by semen from superior bulls. The resulting embryos can be frozen and later inserted in surrogate cows. By circumventing natural reproduction in the interests of developing highly productive herds, we run the risk of developing millions of head of cattle with a very similar genetic makeup. But frozen embryo technology can also make it easier to conserve breeds that are becoming increasingly rare, such as Dutch Belted dairy cattle, Hereford hogs, Lincoln sheep, and Belgian draft horses. Conservation of livestock gene pools warrants immediate attention before more breeds are endangered.

The genetic diversity of forest species must also be conserved in gene banks and natural preserves to ensure reliable supplies of timber and fuelwood for future generations. In the United States, timber species are basically native, and plenty of wild germplasm remains secure *in situ* (Krugman, 1985). Forest-tree improvement programs are relatively young, mostly established during the last thirty years, and most of the tree plantings are genetically similar to their wild counterparts. Because of the rich genetic base of forest trees in the United States, tree plantings based on this germplasm are doing well in over eighty countries. The situation of tropical trees used for stabilizing slopes and providing fuelwood, timber, and fodder has become less favorable as a result of rampant clearing, overgrazing, and excessive fuelwood gathering.

Brazil's genetic resources program has pioneered a seed bank for medicinal plants with a view to improving the yield of useful products from plants for the pharmaceutical industry (CENARGEN, 1984). Collections of some microorganisms are being maintained by Microbiological Resource Centers (MIRCENS), a network of research centers in various regions supported by the United Nations Environment Program (UNEP).

CAREERS IN GENE BANKING

One problem that must be addressed is the staffing of gene banks, particularly in developing countries. The dearth of good gene-bank specialists is not unique to the Third World, however. Gene-bank operators in industrial nations also encounter difficulties in obtaining competent staff. To help further plant genetic resource work in developing countries, IBPGR has sponsored postgraduate training for hundreds of scientists from developing countries. The International Rice Research Institute (IRRI) has recently started a training course for rice germplasm specialists. Similar courses are likely to be established in the near future to cover other crops.

At present gene-bank work is not perceived to be as rewarding as some other branches of science. Germplasm work is not highly visible and some consider it too routine. In many cases the work is seen only as a mundane service to breeders, rather than as a field full of important and challenging problems. If outstanding young people are to make their careers in gene banks, rewards must be more tangible. One way to enhance the profile of gene-bank work, and ultimately recruit bright minds to further the cause of germplasm conservation and utilization, is to encourage gene-bank scientists to focus on critical problems and to publish their results in specialist and interdisciplinary journals.

FINANCIAL SECURITY FOR GENE BANKS

A major worry for most gene-bank curators is funding stability. Electrical costs for cold storage alone can be prohibitive, and all gene banks struggle to meet the costs of collecting, evaluating, and multiplying their accessions. Gene banks, like libraries, require constant support to maintain collections, and collecting gaps are often due to inadequate funding.

How much funding would be required to ensure financial security for gene banks is uncertain, but the estimated $55 million spent in 1982 (Plucknett et al., 1983) is probably close to the amount spent in 1985. Much more evaluation is needed, and expenses for multiplying accessions as well as packaging and shipping them to satisfy requests will undoubtedly increase. In order for gene banks to accomodate the increased work load, between $100 and $150 million per year will be required in the near future to help them function smoothly.

A well-endowed international germplasm fund would be a secure mechanism for supporting the increased need for the conservation and utilization of crop genetic resources. Such a fund should be governed by scientists who understand needs and priorities and who would not propose the use of funds for political purposes. Such a body should also remain in touch with interface developments in biotechology. To make a lasting impact on germplasm conservation and use, the fund's endowment should be sufficiently large to provide between $50 and $75 million a year for disbursements to gene banks. The budget should also be sufficiently flexible to respond to emergency situations.

It is conceivable that a sizable endowment, or at least strong commitments to provide annual funding, could be obtained from private and public sources and placed in trust for that specific purpose. The total required to ensure a minimum level of gene-bank security is relatively small compared to total expenditures on agricultural research worldwide. The developing countries alone spend $1.3 billion a year on agricultural research (Picciotto, 1985).

Although most of the support thus far for germplasm work has come from public sources, the private sector is showing some interest in improving germplasm conservation and evaluation. Pioneer Hi-Bred, for example, has granted $1.5 million, to be released over a five-year period, to the U.S. Department of Agriculture for germplasm research on maize. Pioneer's support of research is exemplary, and we hope that other private firms will follow suit.

GENE BANKS IN THE TWENTY-FIRST CENTURY

By the year 2000, the work of gene banks is likely to have advanced in a number of important areas. Over 90 percent of the remaining variation of the major crops should have been collected, stored, and evaluated by that time. Breeders should have access to computerized catalogs that describe accessions and indicate at least some of their most important traits. Germplasm enhancement should have identified sources of useful genes and many should have been incorporated into parental lines. Furthermore, parental lines will have been crossed sufficiently to establish their inheritance patterns.

Most close wild relatives of crops will have been adequately collected, and widecrossing will be far more commonly used as sources of fresh genes for resistance to insects, diseases, and environmental stresses. Moreover, as recombinant DNA techniques become more sophisticated and reliable, the boundary encompassing close relatives of crops will become ever wider, and collecting for wild species will continue well beyond that for landraces. Some gene banks will have gene libraries from which genes will be easily retrievable for insertion in breeding material. The Japanese Ministry of Agriculture, for example, has earmarked $530,000 to decipher gene sequences in the Ministry's seed stocks and to set up a gene library (Witt, 1985:64). Biotechnology will provide major new tools for plant breeders who will use them to hurdle certain barriers to make genetic improvements.

Gene banks will include an international amalgam of complementary *ex situ* facilities, including tissue-culture collections, for primary germplasm of most crops. Some wild relatives will be stored in *ex situ* gene banks, but many will be maintained in protected natural areas that have

195

been studied extensively for distribution and variability of target populations.

By the year 2000, progress in gene-bank research will have clarified suitable conservation conditions for most crops. Collecting and handling methods for some vegetatively propagated crops will be improved, and we should know what role cryopreservation can play in germplasm storage. And by the end of this century, most countries with either important germplasm resources or plant-breeding programs, or both, will have at least some capacity to store and handle plant germplasm.

Gene-bank development, a product of this century, has come a long way, but much remains to be done. How the world meets the challenge of conserving and using genetic resources will determine how much of the genetic diversity of plants and other organisms survives into the next century.

APPENDICES

APPENDIX 1
International agricultural research centers within the Consultative Group on International Agricultural Research, listed in order of year established.

Acronym	Center	Year Established	Research Programs
IRRI	International Rice Research Institute	1960	Rice
CIMMYT	International Maize and Wheat Improvement Center	1964	Maize, wheat, triticale, barley
IITA	International Institute of Tropical Agriculture	1965	Maize, rice, cowpea, sweet potato, yams, cassava
CIAT	Centro Internacional de Agricultura Tropical	1968	Cassava, beans, rice, pastures
WARDA	West Africa Rice Development Association	1971	Rice
CIP	International Potato Center	1972	Potato
ICRISAT	International Crops Research Institute for the Semi-Arid Tropics	1972	Chickpea, pigeonpea, pear millet, sorghum, groundnut
ILRAD	International Lab. for Research on Animal Diseases	1974	Trypanosomiasis, theileriosis
IBPGR	International Board for Plant Genetic Resources	1973	Plant genetic resources
ILCA	International Livestock Center for Africa	1974	Livestock production systems
IFPRI	International Food Policy Research Institute	1975	Food policy
ICARDA	International Center for Agricultural Research in the Dry Areas	1976	Wheat, barley, triticale, faba bean, lentil, chickpea, forages
ISNAR	International Service for National Agricultural Research	1980	National agricultural research

Notes: Many of the centers also have economic or farming-systems research programs. For locations of centers, see Figure 2.5.

APPENDIX 2

Acronyms and abbreviations for international, regional, and national institutions (public and private) with major crop gene banks.

Acronym	Institution and Location
AES	Agricultural Experiment Station, Australia
AES	Agricultural Experiment Station, Republic of Korea
AICMIP	All India Coordinated Millet Improvement Programme, India
ARARI	Aegean Regional Agricultural Research Organization, Turkey
ARI	Agricultural Research Institute, Burma
ARO	Agricultural Research Organization, Israel
ASP	American Sorghum Project, Yemen Arab Republic
AVRDC	Asian Vegetable Research and Development Center, Taiwan, China
BGC	Barley Germplasm Center, Japan
BRRI	Bangladesh Rice Research Institute, Bangladesh
CRI	Central Agricultural Research Institute, Sri Lanka
CARS	Chitedze Agricultural Research Station, Malawi
CATIE	Centro Agronomico Tropical de Investigación y Enseñanza, Costa Rica
CENARGEN	Centro Nacional de Recursos Genéticos, Brazil
CGI	Crop Germplasm Institute, China
CGR	Cotton Genetic Research, USA
CIAT	Centro Internacional de Agricultura Tropical, Colombia
CICR	Central Institute of Cotton Research, India
CIFEP	Centro de Investigaciónes Fitotécnicas y Ecogenéticas de Pairumani, Bolivia
CIMMYT	Centro Internacional de Mejoramiento de Maiz y Trigo, Mexico
CIP	Centro Internacional de la Papa, Peru
CNIA	Centro Nacional de Investigaciónes Agropecuárias, Argentina
CNPSD	Centro Nacional de Pesquisa de Seringueira e Dendê, Brazil
CNPT	Centro Nacional de Pesquisa de Trigo, Brazil
CNU	Choong-Nam National University, Republic of Korea
CPRI	Central Potato Research Institute, India
CRIFC	Central Research Institute for Food Crops, Indonesia
CRIG	Cocoa Research Institute of Ghana, Ghana
CRRI	Central Rice Research Institute, India
CRU	Cacao Research Unit, Trinidad and Tobago
CSIRO	Commonwealth Scientific and Industrial Research Organization, Australia

CTC	Centro de Tecnologia Copersucar, Brazil
CTCRI	Central Tuber Crops Research Institute, India
CU	Cambridge University, UK
DARS	Darul Aman Research Station, Afghanistan
DGRST	Delegation Generale à la Recherche Scientifique et Technique, Cameroon
EMBRAPA	Empresa Brasileira de Pesquisa Agropecuária, Brazil
FAL	Institut für Pflanzenbau und Pflanzenzuchtung, F.R. Germany
FONAIAP	Fondo Nacional de Investigación y Promoción Apropecuária, Venezuela
FSAE	Faculté des Sciences Agronomique de l'État, Belgium
FSC	Fiji Sugar Corporation, Fiji
FTRS	Fruit Tree Research Station, Japan
GGB	Greek Gene Bank, Greece
HSPA	Hawaiian Sugar Planters' Association, USA
IAC	Instituto Agronômico Campinas, Brazil
IAR	Institute of Agricultural Research, Ethiopia
IARI	Indian Agricultural Research Institute, India
IBP	Institute of Plant Breeding, UK
ICA	Instituto Colombiano Agropecuário, Colombia
ICARDA	International Center for Agricultural Research in the Dry Areas, Syria
ICCPT	National Institute for Cereal and Industrial Crops, Romania
ICRISAT	International Crops Research Institute for the Semi-Arid Tropics, India
IG	Istituto del Germoplasma, Italy
IGPB	Institute of Genetics and Plant Breeding, Czechoslovakia
IHAR	Plant Breeding and Acclimatization Institute, Poland
IIHR	Indian Institute of Horticultural Research, India
IITA	International Institute of Tropical Agriculture, Nigeria
IMPA	Instituto para el Mejoramiento de la Produccion de Azucar, Mexico
IMR	Institute of Maize Research, Yugoslavia
INIA	Instituto Nacional de Investigaciónes Agrarias, Spain
INIA	Instituto Nacional de Investigaciónes Agrícolas, Mexico
INICA	Instituto Nacional de Investigaciónes de la Caña, Cuba
INIPA	Instituto Nacional de Investigación y Promoción Agropecuária, Peru
INRA	Institut National de la Recherche Agronomique, France
INTA	Instituto Nacional de Tecnologia Agropecuária, Argentina
IPB	Institute of Plant Breeding, Finland

IPB	Institute of Plant Breeding, Philippines
IPIGR	Institute of Plant Introduction and Genetic Resources, Bulgaria
IRA	Institut de Recherches Agronomique, Madagascar
IRAT	Institut de Recherches Agronomique Tropicales, France
IRCA	Institut de Recherches sur le Caoutchouc, France
IRCC	Institut de Recherches du Café et du Cacao, France
IRCT	Institut de Recherches du Coton et des Textiles, France
IRFA	Institut de Recherches sur les Fruits et Agrumes, France
IRPIS	Inter-Regional Potato Introduction Station, USA
IRRI	International Rice Research Institute, Philippines
ISRA	Institut Senegalais de Recherches Agricoles, Senegal
ISU	Iowa State University, USA
IVT	Institute for Horticultural Plant Breeding, Netherlands
JII	John Innes Institute, UK
JNRC	Joseph Nickerson Research Center, UK
KNAES	Kyushu National Agricultural Experiment Station, Japan
KSB	Koitotron Seed Bank, Malaysia
KU	Kasetsart University, Thailand
LAA	Liaoning Agricultural Academy, China
MI	Maize Institute, Portugal
MITA	Mayaguez Institute of Tropical Agriculture, Puerto Rico
MRI	Maize Research Institute, Czechoslovakia
MSES	Meringa Sugar Experiment Station, Australia
MU	Malawi University, Malawi
NARS	National Agricultural Research Station, Kenya
NBI	National Biological Institute, Indonesia
NBPGR	National Bureau of Plant Genetic Resources, India
NCSU	North Carolina State University, USA
NGB	Nordic Gene Bank, serving five Nordic countries
NIAR	National Institute of Agricultural Research, Rwanda
NIAS	National Institute of Agricultural Sciences, Japan
NIAVT	National Institute for Agricultural Variety Testing, Hungary
NRCG	National Research Center for Groundnut, India
NRIT	National Research Institute of Tea, Japan
NRPIS	Northeastern Regional Plant Introduction Station, USA
NSSL	National Seed Storage Laboratory, USA
NSWDA	New South Wales Department of Agriculture, Australia
NU	Nairobi University, Kenya
NVRS	National Vegetable Research Station, UK
OBCI	Oil Bearing Crops Institute, China
ORSTOM	Office de la Recherche Scientifique Outre-Mer, France
OSU	Oklahoma State University, USA

PARC	Pakistan Agricultural Research Council, Pakistan
PAU	Punjab Agricultural University, India
PBI	Plant Breeding Institute, UK
PCARRD	Philippine Council for Agricultural Research and Resources Development, Philippines
PGI	Plant Germplasm Institute, Japan
PGRC	Plant Genetic Resources Center, Ethiopia
PGRO	Plant Gene Resources Office, Canada
PORIM	Palm Oil Research Institute of Malaysia
RBG	Royal Botanic Gardens, UK
RICTP	Research Institute for Cereals and Technical Plants, Romania
RIPP	Research Institute of Plant Production, Czechoslovakia
RRI	Rice Research Institute, Thailand
RRIM	Rubber Research Institute of Malaysia, Malaysia
SAA	Shadong Agricultural Academy, China
SBI	Sugarcane Breeding Institute, India
SCRI	Scottish Crop Research Institute, UK
SFS	Sugarcane Field Station, USA
SIRI	Sugarcane Industry Research Institute, Jamaica
SPA	Shensi Province Academy, China
SRPIS	Southern Regional Plant Introduction Station, USA
SVP	Foundation for Agricultural Plant Breeding, Netherlands
TAMU	Texas Agricultural and Mining University, USA
TARI	Taiwan Agricultural Research Institute, Taiwan, China
THRS	Thike Horticultural Research Station, Kenya
TISTR	Thailand Institute of Scientific and Technological Research, Thailand
TRIEA	Tea Research Institute of East Africa, Kenya
TRFCA	Tea Research Foundation of Central Africa, Malawi
TU	Tohoku University, Japan
UB	University of Birmingham, UK
UC	University of California, USA
UF	University of Florida, USA
UM	University of Missouri, USA
UNA	Universidad Nacional Agraria La Molina, Peru
UP	Universidad Politécnica, Spain
UPLB	University of the Philippines at Los Baños, Philippines
USDA	United States Department of Agriculture, USA
USSCFS	United States Sugar Crops Field Station, USA
UWI	University of the West Indies, Trinidad and Tobago
VIR	Vavilov All-Union Institute of Plant Industry, USSR
VPI	Virginia Polytechnic Institute, USA
WARDA	West Africa Rice Development Association, Liberia

WICSCBS		West Indies Central Sugar Cane Breeding Station, Barbados
WRPIS		Western Regional Plant Introduction Station, USA
WRPIS		Western Regional Plant Introduction Station, USA
ZGK		Zentralinstitut für Genetik und Kulturpflanzenforschung, German D.R.

APPENDIX 3
Countries with germplasm storage facilities functioning or under construction as of March 1985.

Country	Storage Facilities Long-term	Storage Facilities Short Medium-term	Country	Storage Facilities Long-term	Storage Facilities Short Medium-term
Afghanistan	No	Yes	Israel	Yes	Yes
Algeria*	Yes	Yes	Italy	Yes	Yes
Argentina	Yes	Yes	Ivory Coast*	Yes	Yes
Australia	Yes	Yes	Japan	Yes	Yes
Austria	Yes	Yes	Kenya*	Yes	Yes
Bangladesh	Yes	Yes	Korea, South	Yes	Yes
Belgium	Yes	Yes	Malawi	No	Yes
Bolivia*	No	Yes	Malaysia*	Yes	Yes
Brazil	Yes	Yes	Mauritius*	No	Yes
Bulgaria	Yes	Yes	Mexico	Yes	Yes
Burkina Faso	No	Yes	Monaco*	Yes	Yes
Canada	Yes	Yes	Mozambique	Yes	Yes
Chile	Yes	Yes	Netherlands	Yes	Yes
China*	Yes	Yes	Niger	Yes	Yes
Colombia	Yes	Yes	Nigeria	Yes	Yes
Costa Rica	Yes	Yes	Nordic (Iceland,		
Cuba*	Yes	Yes	Finland, Norway,		
Cyprus	No	Yes	Sweden, Denmark)	Yes	Yes
Czechoslovakia	No	Yes	Pakistan	No	Yes
Ecuador	No	Yes	Papua New Guinea*	No	Yes
Egypt	No	Yes	Paraguay	No	Yes
Ethiopia	Yes	Yes	Peru	Yes	Yes
Fiji	Yes	Yes	Philippines	Yes	Yes
France	No	Yes	Poland*	Yes	Yes
Germany (D.R.)	Yes	Yes	Portugal*	Yes	Yes
Germany (F.R.)	Yes	Yes	Solomon Islands	No	Yes
Ghana	Yes	Yes	South Africa	Yes	Yes
Greece	Yes	Yes	Spain	Yes	Yes
Hungary	Yes	Yes	Sudan*	Yes	Yes
India	No	Yes	Switzerland	No	Yes
Indonesia	Yes	Yes	Syria	No	Yes
Iran*	Yes	Yes	Thailand	Yes	Yes
Iraq	No	Yes	Togo	Yes	Yes

Tunisia*	Yes	Yes	USA	Yes	Yes
Turkey	Yes	Yes	USSR	Yes	Yes
Uganda	No	Yes	Zambia*	No	Yes
UK	Yes	Yes	Zimbabwe*	Yes	Yes

Note: International agricultural research centers, most of which are listed in Appendix 1, are excluded.

* One or more germplasm storage facilities under construction in 1985.

APPENDIX 4
Base collections of seed crops within the network of the International Board for Plant Genetic Resources, March 1985.

Crop	Institution and Location
CEREALS	
Barley	PGRO, Ottawa, Canada
	NGB, Lund, Sweden
	PGRC, Addis Ababa, Ethiopia
	NIAS, Tsukuba, Japan
Maize	NSSL, Fort Collins, Colo., USA
	NIAS, Tsukuba, Japan
	TISTR, Bangkok, Thailand
	VIR, Leningrad, USSR
	Portuguese Gene Bank, Braga
Millets	PGRC, Addis Ababa, Ethiopia
Eleusine spp.	ICRISAT, Hyderabad, India
Eragrostis spp.	PGRC, Addis Ababa, Ethiopia
Minor Indian millets	IARI, New Delhi, India
Panicum miliaceum	ICRISAT, Hyderabad, India
Pennisetum spp.	PGRO, Ottawa, Canada
	ICRISAT, Hyderabad, India
	NSSL, Fort Collins, Colo., USA
Setaria italica	ICRISAT, Hyderabad, India
Oat	PGRO, Ottawa, Canada
	NGB, Lund, Sweden
Rice	
African species	IITA, Ibadan, Nigeria
indica, javanica	IRRI, Los Baños, Philippines
Mediterranean forms	NSSL, Fort Collins, Colo.,USA
sinica (japonica)	NIAS, Tsukuba, Japan
Wild species	IRRI, Los Banõs, Philippines
Rye	Polish Gene Bank, Radzikow
	NGB, Lund, Sweden
Sorghum	NSSL, Fort Collins, Colo., USA
	ICRISAT, Hyderabad, India

Wheat	VIR, Leningrad, USSR
	NSSL, Fort Collins, Colo.,USA
	IG, Bari, Italy
Wild *Triticum* and *Aegilops* spp.	PGI, Univ. Kyoto, Japan

FOOD LEGUMES

Chickpea	ICRISAT, Hyderabad, India
Faba bean	IG, Bari, Italy
Groundnut	ICRISAT, Hyderabad, India
	INTA, Pergamino, Argentina
Wild perennial spp.	CENARGEN, Brasília, Brazil
Lupin	ZKG, Gatersleben, German D.R.
	INIA, Madrid, Spain
Pea	IG, Bari, Italy
	Polish Gene Bank, Radzikow
Phaseolus	
Wild species	FSAE, Gembloux, Belgium
Cultivated species	CIAT, Cali, Colombia
	NSSL, Fort Collins, Colo., USA
	FAL, Braunschweig, F.R. Germany
Pigeonpea	ICRISAT, Hyderabad, India
Soybean	NSSL, Fort Collins, Colo., USA
Wild perennial spp.	CSIRO, Canberra, Australia
Vigna	
Wild species	FSAE, Gembloux, Belgium
V. radiata	IPB, Los Baños, Philippines
V. Radita	AVRDC, Taiwan, China
V. unguiculata	IITA, Ibadan, Nigeria
	NSSL, Fort Collins, Colo., USA
Winged bean	IPB, Los Baños, Philippines
	TISTR, Bangkok, Thailand

ROOT CROPS

Cassava	CIAT, Cali, Colombia
Potato	CIP, Lima, Peru
Sweet potato	NSSL, Fort Collins, Colo., USA
	AVRDC, Taiwan, China
	NIAS, Tsukuba, Japan

VEGETABLES, OILSEEDS, GREEN MANURES, FODDER

Allium	NVRS, Wellesbourne, UK
	NSSL, Fort Collins, Colo., USA
	NIAVT, Tapioszele, Hungary
	NIAS, Tsukuba, Japan

Amaranths	NSSL, Fort Collins, Colo., USA
	IARI, New Delhi, India
Capsicum	CATIE, Turrialba, Costa Rica
	IVT, Wageningen, Netherlands
Crucifers	
Brassica carinata	PGRC, Addis Ababa, Ethiopia
	FAL, Braunschweig, F.R. Germany
B. oleracea	NVRS, Wellesbourne, UK
	IVT, Wageningen, Netherlands
Oilseeds and green manures	
B. campestris	FAL, Braunschweig, F.R. Germany
B. juncea, B. napus, and	
Sinapis alba	*PGRO*, Ottawa, Canada
Vegetables and fodders	
B. campestris	NVRS, Wellesbourne, UK
B. juncea, B. napus	FAL, Braunschweig, F.R. Germany
All cruciferous crops	NIAS, Tsukuba, Japan
Cabbage	IVT, Wageningen, Netherlands
	NVRS, Wellesbourne, UK
	FAL, Braunschweig, F.R. Germany
Radish	NVRS, Wellesbourne, UK
Wild species	TU, Sendai, Japan
	UP, Madrid, Spain
Eggplant	IVT, Wageningen, Netherlands
	NSSL, Fort Collins, Colo., USA
Okra	NSSL, Fort Collins, Colo., USA
Squashes	
Spp. of *Benincasa, Luffa,*	IPB, Los Baños, Philippines
Momordica, and *Trichosanthes*	
Spp. of *Cucumis* and	NSSL, Fort Collins, Colo., USA
Citrullus	INIA, Madrid, Spain
Cucurbita spp.	NSSL, Fort Collins, Colo., USA
Tomato	CATIE, Turrialba, Costa Rica
	ZGK, Gatersleben, German D.R.
	NSSL, Fort Collins, Colo., USA
	IPB, Los Baños, Philippines
Vegetables	
S.E. Asian species and forms	IPB, Los Baños, Philippines
INDUSTRIAL CROPS	
Beet	FAL, Braunschweig, F.R. Germany
	NGB, Lund, Sweden
	GGB, Thessaloniki, Greece

Cotton	GGB, Thessaloniki, Greece
Sugar cane seed	NSSL, Fort Collins, Colo., USA
Tobacco	GGB, Thessaloniki, Greece
Trees	RBG, Kew, UK
Fuelwood and environmental stabilization	

Note: See Appendix 2 for acronyms.

APPENDIX 5
Field gene banks within the network of the International Board for Plant Genetic Resources, March 1985.

Crop	Institution and Location
ROOTS AND TUBERS	
Cassava	CIAT, Cali, Colombia
	CENARGEN, Brazil*
	INIA, Mexico
	IITA, Ibadan, Nigeria*
Sweet potato	AVRDC, Taiwan, China
FRUITS	
Banana	Banana Board, Jamaica*
	PCARRD, Philippines
	DGRST, Cameroon
Citrus	FTRS, Tsukuba, Japan*
	INIA, Valencia, Spain
	IRFA, Corsica, France*
	USDA**
	CENARGEN, Brazil*
	IIHR, India*
INDUSTRIAL CROPS	
Cacao	UWI, Trinidad and Tobago
	CATIE, Turrialba, Costa Rica
Sugar cane	SBI, Coimbatore, India*
	USDA, Florida, USA
PERENNIALS	
Allium	
Short-day material	Hebrew University, Israel*
Long-day material	RIPP, Olomouc, Czechoslovakia
Groundnut	
Wild perennials	CENARGEN, Brazil

206

Soybean
    Wild perennials   CSIRO, Australia

Note: See Appendix 2 for acronyms of institutions with crop germplasm collections.
* Under discussion or awaiting formal agreement.
** Location being decided.

LITERATURE CITED

Adams, M. W., A. H. Ellingboe, and E. C. Rossman. 1971. Biological uniformity and disease epidemics. *Bioscience* 21:1067-1070.

Allard, R. W. 1970. Problems of maintenance. *In*: O. H. Frankel, E. Bennett, R. D. Brock, A. H. Bunting, J. R. Harlan, and E. Schreiner (eds.), *Genetic Resources in Plants—Their Exploration and Conservation*. F. A. Davis, Philadelphia, pp. 491-494.

Arnold, M. H., D. Astley, E. A. Bell, J.K.A. Bleadsdale, A. H. Bunting, J. Burtley, J. A. Callow, J. P. Cooper, P. R. Day, R. H. Ellis, B. V. Ford-Lloyd, R. J. Giles, J. G. Hawkes, J. D. Hayes, G. G. Henshaw, J. Heslop-Harrison, V. H. Heywood, N. L. Innes, M. T. Jackson, G. Jenkins, M. J. Lawrence, R. N. Lester, P. Matthews, P. M. Mumford, E. H. Roberts, N. W. Simmonds, J. Smartt, R. D. Smith, B. Tyler, R. Watkins, T. C. Whitmore, L. A. Withers. 1986. Plant gene conservation. *Nature* 319:615.

Ashton, P. S. 1981. Tropical botanical gardens: meeting the challenge of declining resources. *Longwood Program Seminar* 13:55-57.

Ayensu, E. E. 1978. The role of science and technology in the economic development of Ghana. *In*: W. Beranek and G. Ranis (eds.), *Science, Technology, and Economic Development: A Historical and Comparative Study*. Praeger, New York, pp. 288-340.

Bajaj, Y.P.S. 1979. Technology and prospects of cryopreservation of germplasm. *Euphytica* 28:267-285.

———. 1983. Cryopreservation and international exchange of germplasm. *In*: S. K. Sin and K. L. Giles (eds.), *Plant Cell Culture in Crop Improvement*. Plenum, New York, pp. 19-41.

Baker, H. G. 1970a. *Plants and Civilization*. Wadsworth Publishing Company, Belmont, California.

———. 1970b. Taxonomy and the biological species concept in cultivated plants. *In*: O. H. Frankel and E. Bennett (eds.), *Genetic Resources in Plants—Their Exploration and Conservation*. Blackwell Scientific Publications, Oxford, pp. 49-68.

———. 1971. Human influences on plant evolution. *Bioscience* 21:108.

Bass, L. N. 1984. Germplasm preservation. *In*: W. L. Brown, T. T. Chang, M. M. Goodman, and Q. Jones (eds.), *Conservation of Crop Germplasm—An International Perspective*. Crop Science Society of America, Madison, Wisconsin, pp. 55-67.

Beardsley, T. 1985. US patent rights march on. *Nature* 317:568.

Beck, B. D. 1982. Historical perspectives of cassava breeding in Africa. *In: Root*

Crops in Eastern Africa: Proceedings of a Workshop Held in Kigali, Rwanda, 23-27 November 1980. International Development Research Centre, Ottawa, pp. 13-18.

Becwar, M. R., P. C. Stanwood, and K. W. Leonhardt. 1983. Dehydration effects on freezing characteristics and survival in liquid nitrogen of desiccation-tolerant and desiccation-sensitive seeds. *Journal of the American Society for Horticultural Science* 108:613-618.

Bernard, C. J. 1945. Le jardin botanique de Buitenzorg et les institutions de botanique appliquée aux Indes Néerlandaises. *In:* P. Honig and F. Verdoorn (eds.), *Science and Scientists in the Netherlands Indies*. Board for the Netherlands Indies, Surinam and Curaçao, New York, pp. 10-15.

Bhatti, M. S., M. Akbar, and A. A. Soomro. 1983. Pakistan. *In: 1983 Rice Germplasm Conservation Workshop*. International Rice Research Institute/International Board for Plant Genetic Resources, Los Baños, Philippines, pp. 35-36.

Bogorad, L. 1983. Overview of the potential and prospects in genetic engineering of plants. *In:* L. W. Shemilt (ed.), *Chemistry and World Food Supplies: The New Frontiers, Chemrawn II*. Pergamon Press, Oxford, pp. 553-562.

Bosemark, N. O. 1979. Genetic poverty of the sugarbeet in Europe. *In:* A. C. Zeven and A. M. Van Harten (eds.), *Proceedings of the Conference: Broadening the Genetic Base of Crops, Wageningen, Netherlands, 3-7 July 1978*. PUDOC, Wageningen, pp. 29-35.

Boster, J. 1983. A comparison of the diversity of Jivaroan gardens with that of the tropical forest. *Human Ecology* 11:47-68.

Boyer, J. S. 1982. Plant productivity and environment. *Science* 218:443-448.

Brandes, E. W., and G. B. Sartoris. 1936. Sugarcane: its origin and improvement. *In: Yearbook of Agriculture*. U.S. Department of Agriculture, Washington, D.C., pp. 561-623.

Bretschneider, E. 1935. *History of European Botanical Discoveries in China*. K. F. Koehlers Antiquarium, Leipzig, Vol. 1.

Brewbaker, J. L. 1979. Diseases of maize in the wet lowland tropics and the collapse of the classic Maya civilization. *Economic Botany* 33:101-118.

Brockway, L. H. 1979. *Science and Colonial Expansion: The Role of the British Royal Botanic Gardens*. Academic Press, New York.

Brown, L. R., and E. C. Wolf. 1985. *Reversing Africa's Decline*. Worldwatch Institute, Paper No. 65, Washington, D.C.

Brown, N. J. 1982. Biological diversity: the global challenge. *In: Proceedings of the U.S. Strategy Conference on Biological Diversity, November 16-18, 1981*. Department of State Publication 9262, Washington, D.C., pp. 22-27.

Brown, W. L. 1982. Genetic diversity: serious business for crop protection and maintenance. *In: Proceedings of the U.S. Strategy Conference on Biological Diversity, November 16-18, 1981*. Department of State Publication 9262, Washington, D.C., pp. 13-17.

———. 1984. Conservation of gene resources in the United States. *In:* C. W. Yeatman, D. Kafton, and G. Wilkes (eds.), *Plant Genetic Resources: A Conservation*

Imperative. Westview Press (AAAS Selected Symposium 87), Boulder, Colorado, pp. 31-41.

Browning, J. A. 1974. Relevance of knowledge about natural ecosystems to development of pest management programs for agro-ecosystems. *Proceedings of the American Phytopathological Society* 1:191-199.

Browning, J. A., and K. J. Frey. 1969. Multiline cultivars as a means of disease control. *Annual Review of Phytopathology* 7:355-382.

Burgess, J. 1984. The revolution that failed. *New Scientist* 104(1428):26-29.

Burkill, I. H. 1918. The establishment of the Botanic Gardens, Singapore. *The Gardens' Bulletin* (Singapore) 2(2):55-63.

Candolle, A. de. 1855. *Géographie Botanique Raisonnée: ou, Exposition des Faits Principaux et des Lois Concernant la Distribution Géographique des Plantes de l'Époque Actuelle*. V. Masson, Paris.

———. 1902. *Origin of Cultivated Plants*. Appleton, New York.

Carneiro, R. L. 1983. The cultivation of manioc among the Kuikuru of the Upper Xingu. *In*: R. B. Hames and W. T. Vickers (eds.), *Adaptive Responses of Native Amazonians*. Academic Press, New York, pp. 65-111.

CENARGEN. 1984. Plantas medicinais—retorno às origens. *CENARGEN Informa* 1:5 (Centro Nacional de Recursos Genéticos, Brazil).

Chang, T. T. 1976a. The rice cultures. *Philosophical Transactions of the Royal Society* (London) B275:146-157.

———. 1976b. *Manual on Genetic Conservation of Rice Germ Plasm for Evaluation and Utilization*. International Rice Research Institute, Los Baños, Philippines.

———. 1976c. Exploitation of useful gene pools in rice through conservation and evaluation. *SABRAO Journal* 8(1):11-16 (Society for the Advancement of Breeding Researchers in Asia and Oceania).

———. 1980. The rice genetic resources program of IRRI and its impact on rice improvement. *In: Rice Improvement in China and Other Asian Countries*. International Rice Research Institute/Chinese Academy of Agricultural Sciences, Los Baños, Philippines, pp. 85-105.

———. 1983a. Genetic resources of rice. *Outlook on Agriculture* 12(2):57-62.

———. 1983b. Guidelines for the cold storage of orthodox crop seeds in the humid tropics. *International Board for Plant Genetic Resources, Regional Committee for Southeast Asia Newsletter* 7(2/3):23-26.

———. 1984a. Conservation of rice genetic resources: luxury or necessity? *Science* 224:251-256.

———. 1984b. Genetics. *Asiaweek* 10(49):94.

———. 1984c. The role and experience of an international crop-specific genetic resources center. *In: Conservation of Crop Germplasm—an International Perspective*. Crop Science Society of America, Madison, Wisconsin, pp. 35-45.

———. 1985. Evaluation and documentation of crop germplasm. *Iowa State Journal of Research* 59:379-397.

Chang, T. T., W. L. Brown, J. G. Boonman, J. Sneep, and H. Lamberts. 1979. Crop genetic resources. *In*: J. Sneep and A.J.T. Hendriksen (eds.), *Plant Breeding Perspectives*. PUDOC, Wageningen, pp. 83-103.

211

Chang, T. T., C. R. Adair, and T. H. Johnston. 1982. The conservation and use of rice genetic resources. *Advances in Agronomy* 35:37-91.

Chapman, C.G.D. 1984. On the size of a genebank and genetic variation it contains. *In*: J.H.W. Holden and J. T. Williams (eds.), *Crop Genetic Resources: Conservation and Evaluation*. George Allen and Unwin, London, pp. 102-119.

Chen, Y. 1983. Malaysia. *In: 1983 Rice Germplasm Conservation Workshop*. International Rice Research Institute/International Board for Plant Genetic Resources, Los Baños, Philippines, pp. 33-34.

Chernela, J. M. In press. Classificação e seleção indígena de grupos subspecíficos de *Manihot esculenta* na área do Rio Uaupés no noroeste da Amazônia. *Anais do XXXV Congresso Nacional de Botanica*.

Chitrakon, S., B. Somrith, and C. Sutthi. 1983. Thailand. *In: 1983 Rice Germplasm Conservation Workshop*. International Rice Research Institute/International Board for Plant Genetic Resources, Los Baños, Philippines, pp. 40-41.

CIAT. 1980. *CIAT Report 1980*. Centro Internacional de Agricultura Tropical, Cali, Colombia.

——. 1984a. CIAT enters the biotechnology era. *CIAT International* 4(1):3 (Centro Internacional de Agricultura Tropical, Cali, Colombia).

——. 1984b. *CIAT 1984: A Summary of Major Achievements during the Period 1977-1983*. Centro Internacional de Agricultura Tropical, Cali, Colombia.

——. 1984c. *External Program Review: Rice Program Report*. Centro Internacional de Agricultura Tropical, Cali, Colombia.

——. 1985. *Annual Report: Genetic Resources Unit Highlights*. Centro Internacional de Agricultura Tropical, Cali, Colombia.

CIMMYT. 1984. *CIMMYT 1983 Annual Report*. Centro Internacional de Mejoramiento de Maiz y Trigo, El Batán, Mexico.

CIP. 1984. *Potatoes for the Developing World*. Centro Internacional de la Papa, Lima.

——. 1985. *International Potato Center: Annual Report 1984*. Centro Internacional de la Papa, Lima.

Clawson, D. L. 1985. Harvest security and intraspecific diversity in traditional tropical agriculture. *Economic Botany* 39:56-67.

Cock, J. H. 1985. *Cassava: New Potential for a Neglected Crop*. Westview Press, Boulder, Colorado.

Comeau, A. 1984. Barley yellow dwarf virus resistance in the genus *Avena*. *Euphytica* 33:49-55.

Conway, R., and D. S. McCauley. 1983. Intensifying tropical agriculture: the Indonesian experience. *Nature* 302:288-289.

Cook, S. F., and W. Borah. 1979. *Essays in Population History: Mexico and California*. University of California Press, Berkeley, Vol. 3.

Creech, J. L. 1970. Tactics of exploration and collection. *In*: O. H. Frankel, E. Bennett, R. D. Brock, A. H. Bunting, J. R. Harlan, and E. Schreiner (eds.), *Genetic Resources in Plants—Their Exploration and Conservation*. F. A. Davis, Philadelphia, pp. 221-229.

Crill, P., J. P. Jones, D. S. Burgis, and S. S. Woltz. 1982a. Controlling Fusarium wilt of tomato with resistant varieties. *In: Evolution of the Gene Rotation Concept for Rice Blast Control: A Compilation of 10 Research Papers*. International Rice Research Institute, Los Baños, Philippines, pp. 1-7.

Crill, P., D. S. Burgis, J. P. Jones, and J. Augustine. 1982b. Tomato variety development and multiple disease control with host resistance. *In: Evolution of the Gene Rotation Concept for Rice Blast Control: A Compilation of 10 Research Papers*. International Rice Research Institute, Los Baños, Philippines, pp. 37-66.

Crill, P., F. L. Nuque, B. A. Estrada, and J. M. Bandong. 1982c. The role of varietal resistance in disease management. *In: Evolution of the Gene Rotation Concept for Rice Blast Control: A Compilation of 10 Research Papers*. International Rice Research Institute, Los Baños, Philippines, pp. 103-121.

Crist, R. E. 1971. Migration and population change in the Irish Republic. *American Journal of Economics and Sociology* 30:253-258.

Crosby, A. W. 1972. *The Columbian Exchange: Biological and Cultural Consequences of 1492*. Greenwood Press, Westport, Connecticut.

Crowe, S., S. Haywood, S. Jellicoe, and G. Patterson. 1972. *The Gardens of Mughul India: A History and a Guide*. Thames and Hudson, London.

Cruls, G. 1949. *Aparência do Rio de Janeiro*. Livraria José Olympio, São Paulo.

Cunningham, I. S. 1984. *Frank N. Meyer: Plant Hunter in Asia*. Iowa State University Press, Ames.

Daniels, J., P. Smith, and N. Paton. 1975. The origins of sugarcanes and centres of diversity in *Saccharum*. *In*: J. T. Williams, C. H. Lamoureux, and N. Wulijarni-Soetjipto (eds.), *South East Asian Plant Genetic Resources*. International Board for Plant Genetic Resources, SEAMEO Regional Center for Tropical Biology/BIOTROP, Badan Penelitan dan Pengembangan Pertanian, and Lembaga Biologi Nasional-LIPI, Bogor, Indonesia, pp. 91-107.

Denevan, W. M. 1983. Adaptation, variation, and cultural geography. *Professional Geographer* 35:399-406.

Dionne, L. A. 1963. Studies on the use of *Solanum acaule* as a bridge between *Solanum tuberosum* and species in the series *Bulbocastana, Cardiophylla* and *Pinnatisecta*. *Euphytica* 12:263-269.

Douglas, J. E. 1980. *Successful Seed Programs: A Planning and Management Guide*. Westview Press, Boulder, Colorado.

Doussinault, G., A. Delibes, R. Sanchez-Monge, and F. Garcia-Olmedo. 1983. Transfer of a dominant gene for resistance to eyespot disease from a wild grass to hexaploid wheat. *Nature* 303:698-700.

Dublin, H. J., and S. Rajaram. 1982. The CIMMYT's international approach to breeding disease-resistant wheat. *Plant Disease* 66:967-972.

Dunlap, V. C. 1967. The United Fruit Company and the Lancetilla Experiment Station. *In: Proceedings of the International Symposium on Plant Introduction, Tegucigalpa, Honduras, November 30-December 2, 1966*. Escuela Agrícola Panamericana, Tegucigalpa, Honduras, pp. 33-41.

Duvick, D. N. 1983. Improved conventional strategies and methods for selection and utilization of germplasm. *In*: L. W. Shemilt (ed.), *Chemistry and World Food Supplies: The New Frontiers, Chemrawn II*. Pergamon Press, Oxford.

Duvick, D. N. 1984. Genetic diversity in major farm crops on the farm and in reserve. *Economic Botany* 38:161-178.

Earnest, E. 1940. *John and William Bartram, Botanists and Explorers.* University of Pennsylvania Press, Philadelphia.

Eckholm, E. P. 1982. *Down to Earth: Environment and Human Needs.* W. W. Norton, New York.

Ehrlich, P., and A. Ehrlich. 1983. *Extinction: The Causes and Consequences of the Disappearance of Species.* Ballantine Books, New York.

Esquinas-Alcazar, J. T. 1981. *Genetic Resources of Tomatoes and Wild Relatives.* International Board for Plant Genetic Resources, Rome.

——. 1982. Recursos fitogenéticos de la región Andina (Parte 5). Food and Agriculture Organization/International Board for Plant Genetic Resources, *Plant Genetic Resources Newsletter* 52:31-36.

Evans, L. T. 1975. Impressions of research on agricultural plants in the USSR. *Journal of the Australian Institute of Agricultural Science* 41:147-155.

Evenson, R. W. 1983. Intellectual property rights and agribusiness research and development: implications for the public agricultural research system. *Amer. J. Agr. Econ.* 65:967-975.

Eyre, A. 1966. *The Botanic Gardens of Jamaica.* Andre Deutsch, London.

Fairchild, D. 1938. *The World Was My Garden: Travels of a Plant Explorer.* Charles Scribner's Sons, New York.

FAO. 1983a. *Plant Genetic Resources: Report of the Director-General.* Food and Agriculture Organization of the United Nations, Rome.

——. 1983b. *Report of the Conference of FAO, Twenty-second Session, Rome, 5-23 November 1983.* Food and Agriculture Organization of the United Nations, Rome.

——. 1985. *Information Note on the Seed Improvement and Development Programme.* AGP/SIDP/85/15, Food and Agriculture Organization of the United Nations, Rome.

Faris, D. G. 1984. ICRISAT's research on pigeonpea. *In: Grain Legumes in Asia.* International Crops Research Institute for the Semi-Arid Tropics, Patancheru, India, pp. 17-20.

Feistritzer, W. P. 1975. *General Seed Technology.* Food and Agriculture Organization of the United Nations, Rome.

Feldman, M. 1976. Wheats. *In:* N. W. Simmonds (ed.), *Evolution of Crop Plants.* Longman, London, pp. 120-128.

Ferrez, G., and M. Mouillot. 1965. *A Muito Leal e Heróica Cidade de São Sebastião do Rio de Janeiro.* Raymundo de Castro Maya, Candido Guinle de Paula Machado, Fernando Machado Portella, Banco Boavista S.A., Rio de Janeiro.

Fischbeck, G. 1981. The usefulness of gene banks: perspectives for the breeding of plants. *In: The Use of Genetic Resources in the Plant Kingdom.* Union Internationale pour la Protection des Obtentions Végétales, Geneva, pp. 15-26.

Ford-Lloyd, B., and M. Jackson. 1984. Plant gene banks at risk. *Nature* 308:683.

Foster, B. 1984. Native plant gene conservation in British Columbia. *In:* C. W. Yeatman, D. Kafton, and G. Wilkes (eds.), *Plant Genetic Resources: A Conserva-*

tion Imperative. Westview Press (AAAS Selected Symposia Series 87), Boulder, Colorado, pp. 63-70.

Frankel, O. H. 1970. Genetic conservation in perspective. *In*: O. H. Frankel, E. Bennett, R. D. Block, A. H. Bunting, J. R. Harlan, and E. Schreiner (eds.), *Genetic Resources in Plants—Their Exploration and Conservation*. F. A. Davis, Philadelphia, pp. 469-489.

———. 1975. Genetic conservation—why and how. *In*: J. T. Williams, C. H. Lamoureux, and N. Wulijarni-Soetjipto (eds.), *South East Asian Plant Genetic Resources*. International Board for Plant Genetic Resources, SEAMEO Regional Center for Tropical Biology/BIOTROP, Badan Penelitan dan Pengembangan Pertanian, and Lembaga Biologi Nasional-LIPI, Bogor, Indonesia, pp. 16-32.

———. 1977. Natural variation and its conservation. *In*: A. Muhammed, R. Aksel, and R. C. Von Borstel (eds.), *Genetic Diversity in Plants*. Plenum Press, New York, pp. 21-44.

———. 1981. Conservation of genes, gene banks and patents. *In*: H. Messel (ed.), *The Biological Manipulation of Life*. Pergamon Press, Sydney, p. 212.

Frankel, O. H., and E. Bennett (eds.). 1970. *Genetic Resources in Plants: Their Exploration and Conservation*. Blackwell, Oxford.

Frankel, O. H., and J. G. Hawkes (eds.). 1975. *Crop Genetic Resources for Today and Tomorrow*. Cambridge University Press, Cambridge.

Frankel, O. H., and M. E. Soulé. 1981. *Conservation and Evolution*. Cambridge University Press, Cambridge.

Frey, K. J., J. A. Browning, and M. D. Simons. 1973. Management of host resistance genes to control diseases. *Zeitschrift fur Pflanzenkrankheiten und Pflanzenschutz* 80:160-180.

Fry, W. E. 1982. *Principles of Plant Disease Management*. Academic Press, New York.

Galinat, W. C., P. C. Mangelsdorf, and L. Piersen. 1956. Estimates of teosinte introgression in archaeological maize. *Bot. Mus. Leafl. Harvard University* 17:101-124.

Galloway, J. H. 1985. Tradition and innovation in the American sugar industry, c. 1500-1800: an explanation. *Annals of the Association of American Geographers* 75:334-351.

Gardner, G. 1846. *Travels in the Interior of Brazil, Principally through the Northern Provinces, and the Gold and Diamond Districts, during the Years 1836-1841*. Reeve Brothers, London.

Gill, K. S., G. S. Nanda, and G. Singh. 1984. Stability analysis over seasons and locations of multilines of wheat (*Triticum aestivum* L.). *Euphytica* 33:489-495.

Glacken, C. J. 1976. *Traces on the Rhodian Shore: Nature and Culture in Western Thought from Ancient Times to the End of the Eighteenth Century*. University of California Press, Berkeley.

Godden, D. 1984. Plant breeders' rights and international agricultural research. *Food Policy* 9(3):206-218.

Gomez, P. L. 1985. Mejoramiento genético de la papa. Manuscript.

Goodland, R. 1985. Wildland management in economic development. Keynote address at the First International Wildlife Symposium, IX World Forestry Congress and the Wildlife Society of Mexico, Mexico City, 14 May.

Goodman, M. M. 1976. Maize. *In*: N. W. Simmonds (ed.), *Evolution of Crop Plants*. Longman, London, pp. 128-136.

Grassel, C. O. 1965. Introgression between *Saccharum* and *Miscanthus* in New Guinea and the Pacific area. *Proc. 12th Int. Congr. Sugar Cane Technologists, Puerto Rico*. Elsevier, Amsterdam, pp. 995-1003.

Griliches, Z. 1958. Research costs and social returns: hybrid corn and related innovations. *Journal of Political Economy* 66:419-431.

Hahn, S. K. 1978. Breeding cassava for resistance to bacterial blight. *PANS* 24:480-485.

Hahn, S. K., E. R. Terry, K. Leuschner, I. O. Akobundu, C. Okali, and R. Lal. 1979. Cassava improvement in Africa. *Field Crops Research* 2:193-226.

Hahn, S. K., E. R. Terry, and K. Leuschner. 1980a. Breeding cassava for resistance to cassava mosaic disease. *Euphytica* 29:673-683.

Hahn, S. K., A. K. Howland, and E. R. Terry. 1980b. Correlated resistance of cassava to mosaic and bacterial blight diseases. *Euphytica* 29:305-311.

Hammons, R. O. 1976. Peanuts: genetic vulnerability and breeding strategy. *Crop Science* 16:527-530.

Hanneman, R. E. 1976. The Inter-Regional Potato Introduction Project (IR-1). *In: International Potato Center Report of the Planning Conference on the Exploration and Maintenance of Germ Plasm Resources*. Centro Internacional de la Papa, Lima.

Hanson, J., J. T. Williams, and R. Freund. 1984. *Institutes Conserving Crop Germplasm: The IBPGR Global Network of Genebanks*. International Board for Plant Genetic Resources, Rome.

Hardon, J. J. 1976. Oil palm. *In*: N. W. Simmonds (ed.), *Evolution of Crop Plants*. Longman, London, pp. 225-229.

Hardy, R.W.F. 1983. Chemical, biological, genetic, and agronomic approaches to improved or alternative technologies to provide fixed nitrogen. *In*: L. W. Shemilt (ed.), *Chemistry and World Food Supplies: The New Frontiers, Chemrawn II*. Pergamon Press, Oxford, pp. 585-599.

———. 1984. Biotechnology: status, forecast and issues. Prepared as Technology Sector Forecast for Symposium on Technological Frontiers and Foreign Relations.

Harlan, H. V., and M. L. Martini. 1936. Problems and results in barley breeding. *In: Yearbook of Agriculture*, U.S. Department of Agriculture, Washington, D.C., pp. 303-346.

Harlan, J. R. 1965. The possible role of weed races in the evolution of cultivated plants. *Euphytica* 14:173-176.

———. 1972. Genetics of disaster. *Journal of Environmental Quality* 1:212-215.

———. 1975a. Our vanishing genetic resources. *Science* 188:618-621.

———. 1975b. *Crops and Man*. American Society of Agronomy/Crop Science Society of America, Madison, Wisconsin.

216

——. 1976. Genetic resources in wild relatives of crops. *Crop Science* 16:329-333.

——. 1984. Evaluation of wild relatives of crop plants. *In*: J.H.W. Holden and J. T. Williams (eds.), *Crop Genetic Resources: Conservation and Evaluation.* George Allen and Unwin, London, pp. 212-222.

Harrington, J. F. 1970. Seed and pollen storage for conservation of plant gene resources. In: O. H. Frankel, E. Bennett, R. D. Brock, A. H. Bunting, J. R. Harlan, and E. Schreiner (eds.), Genetic Resources in Plants—Their Exploration and Conservation. F. A. Davis, Philadelphia, pp. 501-521.

Hart, W. E. 1919. The botanic garden of Pamplemousses. *Royal Botanic Gardens, Kew, Bulletin of Miscellaneous Information* 5:279-286.

Hartley, C. 1939. The clonal variety for tree planting: asset or liability? *Phytopathology* 29:9.

Harvey, J. 1981. *Mediaeval Gardens*. B. T. Batsford, London.

Hawkes, J. G. 1958. Significance of wild species and primitive forms for potato breeding. *Euphytica* 7:257-270.

——. 1969. The ecological background of plant domestication. *In*: P. J. Ucko and G. W. Dimbleby (eds.), *The Domestication and Exploitation of Plants and Animals*. Duckworth, London, pp. 17-29.

——. 1977a. Plant gene pools—an essential resource for the future. *Journal of the Royal Society of Arts* 125:224-235.

——. 1977b. The importance of wild germplasm in plant breeding. *Euphytica* 26:615-621.

——. 1978. The taxonomist's role in the conservation of genetic diversity. *In*: H. E. Street (ed.), *Essays in Plant Taxonomy*. Academic Press, New York, pp. 775-783.

——. 1979. Genetic poverty of the potato in Europe. In: A. C. Zeven and A. M. van Harten (eds.), *Proceedings of the Conference: Broadening the Genetic Base of Crops, Wageningen, Netherlands, 3-7 July, 1978*. PUDOC, Wageningen, pp. 19-27.

——. 1980. The taxonomy of cultivated plants and its importance in plant breeding research. *In: Perspectives in World Agriculture*. Commonwealth Agricultural Bureaux, Farnham Royal, Slough, England, pp. 49-66.

——. 1981. Biosystematic studies of cultivated plants as an aid to breeding research and plant breeding. *Kulturpflanze* 29:327-335.

——. 1982. Genetic conservation of "recalcitrant species"—an overview. *In*: L. A. Withers and J. T. Williams (eds.), *Crop Genetic Resources: The Conservation of Difficult Material*. IUBS/IBPGR, Paris, pp. 83-92.

——. 1983. *The Diversity of Crop Plants*. Harvard University Press, Cambridge, Massachusetts.

——. 1985. *Plant Genetic Resources: The Impact of the International Agricultural Research Centers*. Consultative Group on International Agricultural Research, World Bank, Washington, D.C., Study Paper No. 3.

Healey, B. J. 1975. *The Plant Hunters*. Charles Scribner's Sons, New York.

Heinrichs, E. A. 1984. Perspectives and directions for the continued development of insect resistant varieties. Paper presented at the Symposium on Plant

217

Resistance to Insects: Research Strategies for the 21st Century, Annual Meeting of the Entomological Society of America, December 12, 1984, San Antonio, Texas.

Heinz, D. J. 1967. Wild *Saccharum* species for breeding in Hawaii. *In: Proceedings of the Twelfth International Society of Sugarcane Technologists, Puerto Rico, 1965.* Elsevier, Amsterdam, pp. 1037-1043.

Hemming, J. 1978a. *Red Gold: The Conquest of the Brazilian Indians, 1500-1760.* Harvard University Press, Cambridge, Massachusetts.

——. 1978b. *The Search for El Dorado.* Michael Joseph, London.

Hepper, F. N. 1982. *Royal Botanic Gardens, Kew: Gardens for Science and Pleasure.* Her Majesty's Stationary Office, London.

Hester, J. H. 1983. Plant, science, and human needs: the New York Botanical Garden gears up. *Orion* 2(4):24-31.

Hibino, H. 1984. Rice grassy stunt virus: current research and prospects. Paper presented at the Workshop on the RTV Collaborative Project, 20-21 October, Bhubaneswar, Orissa, India.

Hinman, C. W. 1984. New crops for arid lands. *Science* 225:1445-1448.

Holden, J.H.W. 1984. The second ten years. *In*: J.H.W. Holden and J. T. Williams (eds.), *Crop Genetic Resources: Conservation and Evaluation.* George Allen and Unwin, London, pp. 277-285.

Holttum, R. E. 1984. Tropical botanic gardens, past, present and future. Keynote address at the International Symposium on Botanic Gardens of the Tropics, Penang, Malaysia, 17-19 December.

Hooykaas-Van Slogteren, G.M.S., P.J.J. Hooykaas, and R. A. Schilperoort. 1984. Expression of Ti plasmid genes in monocotyledonous plants infected with *Agrobacterium tumefasciens. Nature* 311:763-764.

Howard, R. A. 1954. A history of the botanic garden of St. Vincent, British West Indies. *Geographical Review* 44:381-393.

Huamán, Z. 1982. The world potato collection maintained at CIP. Paper presented at the International Germ Plasm Course, Lima, January 11 to February 19.

Huamán, J. G. Hawkes, and P. R. Rowe. 1980. *Solanum ajanhuiri*: An important diploid potato cultivated in the Andean altiplano. *Economic Botany* 34:335-343.

Humboldt, A. von. 1818. *Personal Narrative of Travels to the Equinoctial Regions during the Years 1799-1804.* Longman, London, Vol. 1.

Hussaini, S. H., M. M. Goodman, and D. H. Timothy. 1977. Multivariate analysis and the geographical distribution of the world collection of finger millet. *Crop Science* 17:257-263.

Hyams, E., and W. MacQuitty. 1969. *Great Botanical Gardens of the World.* MacMillan, London.

Hyland, H. L. 1977. History of U.S. plant introduction. *Environmental Review* 4/77:26-33.

——. 1984. History of plant introduction in the United States. *In*: C. W. Yeatman, D. Kafton, and G. Wilkes (eds.), *Plant Genetic Resources: A Conservation*

Imperative. Westview Press (AAAS Selected Symposia Series 87), Boulder, Colorado, pp. 6-14.

IARI. 1980. *The New Production Technology for Barley*. Indian Agricultural Research Institute, New Delhi, Research Bulletin 24.

IBPGR. 1976. *IBPGR Advisory Committee on Sorghum and Millets Germplasm: Report of the First Meeting held 3-7 October, 1976, ICRISAT, Hyderabad, India*. International Board for Plant Genetic Resources, Rome.

——. 1980. *Directory of Germplasm Collections*. International Board for Plant Genetic Resources, Rome.

——. 1983a. *Practical Considerations Affecting the Collection and Exchange of Samples of Wild Species and Primitive Cultivars*. International Board for Plant Genetic Resources, Rome.

——. 1983b. *IBPGR Advisory Committee on in vitro Storage: Report of the First Meeting*. International Board for Plant Genetic Resources, Rome.

——. 1983c. *Crop Genetic Resources for All*. International Board for Plant Genetic Resources, Rome.

——. 1983d. *Kodo Millet Descriptors*. International Board for Plant Genetic Resources, Rome.

——. 1983e. *Pear Descriptors*. International Board for Plant Genetic Resources/Commission of European Communities, Rome and Brussels.

——. 1984a. Annual Report 1983. International Board for Plant Genetic Resources, Rome.

——. 1984b. *The IBPGR in Its Second Decade: An Updated Strategy and Planning Report*. International Board for Plant Genetic Resources, Rome.

——. 1985a. *IBPGR Advisory Committee on in vitro Storage: Report of the Second Meeting*. International Board for Plant Genetic Resources, Rome.

——. 1985b. *IBPGR Advisory Committee on Seed Storage: Report of the Third Meeting*. International Board for Plant Genetic Resources, Rome.

——. 1985c. *Ecogeographical Surveying and in situ Conservation of Crop Relatives*. International Board for Plant Genetic Resources, Rome.

——. 1985d. *Annual Report 1984*. International Board for Plant Genetic Resources, Rome.

ICARDA. 1984. *A Genetic Resource Program at ICARDA: Strategies and a Five-Year Work Program*. International Center for Agriculture in the Dry Areas, Aleppo, Syria.

ICRISAT. 1978. *ICRISAT Annual Report 1977-1978*. International Crops Research Institute for the Semi-Arid Tropics, Patancheru, India.

——. 1985. *Proceedings of an International Workshop on Cytogenetics of Arachis, ICRISAT Center, Patancheru, India, 31 Oct-2 Nov 1983*. International Crops Research Institute for the Semi-Arid Tropics, Patancheru, India.

IITA. 1983. *IITA Research Highlights '82*. International Institute of Tropical Agriculture, Ibadan, Nigeria.

——. 1985. *IITA Research Highlights 1984*. International Institute of Tropical Agriculture, Ibadan, Nigeria.

Ingram, G. B., and J. T. Williams. 1984. *In situ* conservation of wild relatives of

219

crops. *In*: J.H.W. Holden and J. T. Williams (eds.), *Crop Genetic Resources: Conservation and Evaluation*. George Allen and Unwin, London, pp. 163-179.

Innes, N. L. 1975. Genetic conservation and the breeding of field vegetables for the United Kingdom. *Outlook on Agriculture* 8:301-305.

IRRI. 1980. IRRI germplasm bank—treasure of mankind. *IRRI Reporter* 3/80:2. International Rice Research Institute, Los Baños, Philippines.

———. 1984. *Annual Report for 1983*. International Rice Research Institute, Los Baños, Philippines.

Jain, H. K. 1982. Plant breeders' rights and genetic resources. *Indian Journal of Genetics and Plant Breeding* 42(2):121-128.

Jain, H. K., and S. K. Banerjee. 1982. Problems of seed production and procedures of varietal release in India. *Seed Research* 10:1-17.

Jennings, D. L. 1976. Cassava. *In*: N. W. Simmonds (ed.), *Evolution of Crop Plants*. Longman, London, pp. 81-84.

Johnson, R., and D. J. Allen. 1975. Induced resistance to rust diseases and its possible role in the resistance of multiline varieties. *Annals of Applied Biology* 80:359-363.

Johnson, V. A., and H. L. Beemer. 1977. *Wheat in the Peoples' Republic of China*. National Academy of Sciences, Washington, D.C.

Jones, M.G.K. 1985. Transformation of cereal crops by direct gene transfer. *Nature* 317:579-580.

Jones, W. O. 1959. *Manioc in Africa*. Stanford University Press, Stanford, California.

Kahn, E. J. 1985. *The Staffs of Life*. Little, Brown and Company, Boston.

Kartha, K. K., L. A. Mroginski, K. Pahl, and N. L. Leung. 1981. Germplasm preservation of coffee (*Coffea arabica* L.) by *in vitro* culture of shoot apical meristems. *Plant Sci. Lett.* 22:301-308.

KCLRS. 1983. *Wheat Varieties: Special Press Release*. Kansas Crop and Livestock Reporting Service, Kansas State Board of Agriculture, Topeka, Kansas, 11 February.

Kerr, W. E., and C. R. Clement. 1980. Prácticas agrícolas de consequências genéticas que possibilitaram aos índios da Amazônia uma melhor adaptação as condições ecológicas da região. *Acta Amazonica* 10:251-261.

Kerr, W. E., and D. A. Posey. 1984. Informações adicionais sobre a agricultura dos Kayapó. *Interciencia* 9:392-400.

Khush, G. S. 1977. Disease and insect resistance in rice. *Advances in Agronomy* 29:265-341.

———. In press. IRRI breeding program and its worldwide impact on increasing rice production. *In: Gene Manipulation in Plant Improvement*. Plenum Press, New York.

Khush, G. S., and K. C. Ling. 1974. Inheritance of resistance to grassy stunt virus and its vector in rice. *Journal of Heredity* 65:134-136.

Khush, G. S., K. C. Ling, R. C. Aquino, and V. M. Aguiero. 1977. Breeding for resistance to grassy stunt in rice. Proceedings 3rd International Congress of the Society for the Advancement of Breeding Researchers in Asia and Oceania (SABRAO). *Plant Breeding Papers* 1(4b):3-9.

Kingdon Ward, F. 1924. *The Romance of Plant Hunting*. Edward Arnold, London.

Klose, N. 1950. *America's Crop Heritage: The History of Foreign Plant Introduction by the Federal Government*. Iowa State College Press, Ames.

Knott, D. R., and J. Dvorak. 1976. Alien germ plasm as a source of resistance to disease. *Annual Review of Phytopathology* 14:211-235.

Konzak, C. F. 1984. Role of induced mutations. *In*: P. B. Vose and S. G. Blixt (eds.), *Crop Breeding—A Contemporary Basis*. Pergamon Press, Oxford, pp. 216-292.

Konzak, C. F., A. Kleinhofs, and S. E. Ullrich. 1984. Induced mutations in seed-propagated crops. *In*: J. Janick (ed.), *Plant Breeding Reviews*, Vol. 2. Avi Publishing Co., Westport, Connecticut, pp. 13-72.

Krugman, S. L. 1985. Forest genetics and foreign policy. *Iowa State Journal of Research* 59:529-539.

Kyaw, U. O. 1983. Burma. *In: 1983 Rice Germplasm Conservation Workshop*. International Rice Research Institute/International Board for Plant Genetic Resources, Los Baños, Philippines, pp. 29-30.

Leaf, M. J. 1983. The green revolution and cultural change in a Punjab village, 1965-1978. *Economic Development and Cultural Change* 31:227-270.

Lehman, C. O. 1979. The Gatersleben gene bank. *In*: A. C. Zeven and A. M. Van Harten (eds.), *Proceedings of the Conference: Broadening the Genetic Base of Crops, Wageningen, Netherlands, 3-7 July 1978*. PUDOC, Wageningen, pp. 111-116.

Leuschner, K. 1981. Screening for resistance against the green spider mite. *In*: E. R. Terry, K. A. Oduro, and F. Cavenes (eds.), *Tropical Root Crops: Research Strategies for the 1980s*. International Development Research Centre, Ottawa, pp. 75-78.

——. 1982. Pest control for cassava and sweet potato. *In: Root Crops in Eastern Africa: Proceedings of a Workshop Held in Kigali, Rwanda, 23-27 November 1980*. International Development Research Centre, Ottawa, pp. 60-64.

Lewin, R. 1982. Never-ending race for genetic variants. *Science* 218:877.

Lewis, L. N. 1985. Genetic engineering: one leg of a three-legged stool. *California Agriculture* 39(1 and 2):2.

Lucas, G. B. 1980. The war against blue mold. *Science* 210:147-153.

Luthra, J. K., and M. V. Rao. 1979. Escape mechanism operating in multilines and its significance in relation to leaf rust epidemics. *Indian Journal of Genetics and Plant Breeding* 39:38-49.

Lyman, J. M. 1984. Progress and planning for germplasm conservation of major food crops. *Plant Genetic Resources Newsletter* (Food and Agriculture Organization/International Board for Plant Genetic Resources) 60:3-21.

McAlister, L. N. 1984. *Spain and Portugal in the New World, 1492-1700*. University of Minnesota Press, Minneapolis.

McDonald, D. 1984. ICRISAT's research on groundnut. *In: Grain Legumes in*

221

Asia. International Crops Research Institute for the Semi-Arid Tropics, Patancheru, India, pp. 14-23.

MacFadyen, J. T. 1985. A battle over seeds: the Third World asks for a share of gene stocks bred in northern laboratories—from southern seed. *Atlantic* 256(5):36-44.

Mackenzie, D. 1983. Seeds of conflict over food genes. *New Scientist* 1390:870-871.

McClean, T. 1981. *Medieval English Gardens*. Viking Press, New York.

MacPhail, I. 1972. *Hortus Botanicus: The Botanic Garden and the Book*. Sterling Morton Library/Newberry Library, Lisle, Illinois.

Majid, A., and A. Hendranata. 1975. Selection and conservation problems in Hevea with special reference to Indonesia. *In*: J. T. Williams, C. H. Lamoureux, and N. Wulijarni-Soetjipto (eds.), *South East Asian Plant Genetic Resources*. International Board for Plant Genetic Resources, SEAMEO Regional Center for Tropical Biology/BIOTROP, and Lembaga Biologi Nasional-LIPI, Bogor, Indonesia, pp. 171-177.

Mangelsdorf, P. C., R. S. MacNeish, and W. C. Galinat. 1967. Prehistoric wild and cultivated maize. *In*: D. S. Byers (ed.), *The Prehistory of the Tehuacan Valley, Volume One: Environment and Subsistence*. University of Texas Press, Austin, pp. 178-200.

Martin, J. P. 1965. The commercial sugar cane varieties of the world and their resistance and susceptibility to the major diseases. *Proc. 12th Int. Congress of the Int. Soc. Sugar Cane Technologists, Puerto Rico*. Elsevier, Amsterdam, pp. 1212-1225.

Massart, J. 1945. Notes Javanaises. *In*: P. Honig and F. Verdoorn (eds.), *Science and Scientists in the Netherlands Indies*. Board for the Netherlands Indies, Surinam and Curaçao, New York, pp. 231-240.

May, R. M. 1985. Evolution of pesticide resistance. *Nature* 315:12-13.

Melendez, H., M.A.U. Lopez, J. A. Espinoza, V.W.G. Lauck, F.A.I. Lopez, A. M. Rodriguez, F. O. Chavez, L. P. Aponto, P. R. Beltran, and J. V. Palma. 1981. *Sabanera y Costeña dos Nuevas Variedades de Yuca para el Trópico Humedo de México*. Secretaría de Agricultura y Recursos Hidráulicos, INIA, Centro de Investigaciones Agrícolas del Golfo Centro, Campo Experimental Huimanguillo, Huimanguillo, Tabasco, Mexico, Folleto Técnico 1.

Mengesha, M. H. 1984. International germplasm collection, conservation, and exchange at ICRISAT. *In*: W. L. Brown, T. T. Chang, M. M. Goodman, and Q. Jones (eds.), *Conservation of Crop Germpaslm—An International Perspective*. Crop Science Society of America, Madison, Wisconsin, pp. 47-54.

Miche, A., A. Maluszynksi, and B. Donini. 1985. *Plant Cultivars Derived from Mutation Induction or the Use of Induced Mutations in Cross Breeding*. Mutation Breeding Review No. 3, Food and Agriculture Organization/International Atomic Energy Agency, Vienna.

Mooney, P. R. 1979. *Seeds of the Earth: A Private or Public Resource?* Inter Pares, Ottawa.

———. 1983. *The Law of the Seed: Another Development and Plant Genetic Resources.* Development Dialogue 1983:1-2, Dag Hammarskjold Foundation, Uppsala.

Moss, J. P. 1980. Wild species in the improvement of groundnuts. *In*: R. J. Summerfield and A. H. Bunting (eds.), *Advances in Legume Science*. Royal Botanic Gardens, Kew, pp. 525-535.

Murata, M., E. E. Roos, and T. Tsuchiya. 1981. Chromosome damage induced by artificial seed ageing in barley. *Canadian Journal of Genetics and Cytology* 23:267-280.

Murphy, C. F. 1985. The National Plant Germplasm System (NPGS): Report of Acting Assistant to Deputy Administrator to Regional Technical Committees. Mimeo.

Myers, N. 1983. *A Wealth of Wild Species: Storehouse for Human Welfare*. Westview Press, Boulder, Colorado.

———. 1984. *The Primary Source: Tropical Forests and Our Future*. W. W. Norton, New York.

NAS. 1972. *Genetic Vulnerability of Major Crops*. National Academy of Sciences, Washington, D.C.

NASULGC. 1983. *Emerging Biotechnologies in Agriculture: Issues and Policies*. Division of Agriculture, Committee on Biotechnology, National Association of State Universities and Land-Grant Colleges, Washington, D.C.

Nevo, E., A.H.D. Brown, and D. Zohary. 1979. Genetic diversity in the wild progenitor of barley in Israel. *Experientia* 35:1027-1029.

Ng, N. Q. 1979. Plant exploration in central Nigeria. *In: Genetic Resources Exploration*. International Institute of Tropical Agriculture, Ibadan, Nigeria, pp. 76-81.

Ng, N. Q., M. Jacquot, A. Abifarin, K. Goli, A. Ghesquiere, and K. Miezan. 1983. Rice genetic resources collection and conservation activities in Africa and Latin America: programs of IITA, WARDA, IDESSA, IRAT, and ORSTOM. *In: 1983 Rice Germplasm Conservation Workshop*. International Rice Research Institute/The International Board for Plant Genetic Resources, Los Baños, Philippines, pp. 45-52.

Noorsyamsi, H., and O. O. Hidayat. 1974. The tidal swamp rice culture in South Kalimantan. *Contributions from the Central Research Institute for Agriculture Bogor* 10:1-18.

NRC. 1984. *Amaranth: Modern Prospects for an Ancient Crop*. National Research Council, National Academy Press, Washington, D. C.

Oka, H. I. 1983. Conservation of heterogeneous rice populations. *In: 1983 Rice Germplasm Conservation Workshop*. International Rice Research Institute/The International Board for Plant Genetic Resources, Los Baños, Philippines, pp. 45-52.

Oster, G., and S. Oster. 1985. The great breadfruit scheme: a beautiful tree still bears the stigma of its past. *Natural History* 94(3):35-41.

Palmer, L. T., Y. Soepriaman, and S. Kartaatmadja. 1978. Rice yield losses due to brown planthopper and rice grassy stunt disease in Java and Bali. *Plant Disease Reporter* 62:962-965.

223

Pandey, H. N. 1984. Development of multilines in durum wheat cultivars. *Rachis* 3:9-10.

Patiño, V. M. 1963. *Plantas Cultivadas y Animales Domésticos en América Equinoccial.* Imprenta Departamental, Cali, Colombia, Vol. 2.

Peacock, W. J. 1984. The impact of molecular biology on genetic resources. *In:* J.H.W. Holden and J. T. Williams (eds.), *Crop Genetic Resources: Conservation and Evaluation.* George Allen and Unwin, London, pp. 268-276.

Permar, J. H. 1945. *Catalog of Plants Growing in the Lancetilla Experimental Garden at Tela, Honduras.* Compañia Editora de Honduras, San Pedro Sula, Honduras.

Phillips, L. L. 1976. Cotton. *In:* N. W. Simmonds (ed.), *Evolution of Crop Plants.* Longman, London, pp. 196-200.

Phillips, R. L. 1984. Implications of molecular genetics in plant breeding and opportunities for collaboration. Paper presented at the Plant Breeding Research Forum, Wayzata, Minnesota, August 22-24.

Picciotto, R. 1985. National agricultural research: testing the feasibility of agricultural research schemes in developing nations. *Finance and Development* 22(2):45-48.

Pioneer. 1982. *Plant Breeding Research: An Investment in Food Security (A Summary of Findings of the 1982 Plant Breeding Research Forum Sponsored by Pioneer Hi-Bred International, Inc.).* Pioneer Hi-Bred International, Inc., Des Moines, Iowa.

Plucknett, D. L., and N.J.H. Smith, 1982. Agricultural research and Third World food production. *Science* 217:215-220.

——. 1984. Networking in international agricultural research. *Science* 225:989-993.

——. 1986a. Historical perspectives on multiple cropping. *In:* C. A. Francis (ed.), *Multiple Cropping.* MacMillan, New York, pp. 20-39.

——. 1986b. Sustaining agricultural yields: as productivity rises, maintenance research is needed to uphold the gains. *Bioscience* 36:40-45.

Plucknett, D. L., N.J.H. Smith, J. T. Williams, and N. Murthi Anishetty. 1983. Crop germplasm conservation and developing countries. *Science* 220:163-169.

Popovsky, M. 1984. *The Vavilov Affair.* Archon Books, Hamden, Connecticut.

Porto, P. C. 1936. Plantas indigenas e exoticas provenientes da Amazonia, cultivadas no Jardim Botanico do Rio de Janeiro. *Rodriguesia* 2(5):93-157.

Posey, D. A. 1984. A preliminary report on diversified management of tropical forest by the Kayapó Indians of the Brazilian Amazon. *Advances in Economic Botany* 1:112-126.

——. 1985. Indigenous management of tropical forest ecosystems: the case of the Kayapó Indians of the Brazilian Amazon. *Agroforestry Systems* 3:139-158

Prance, G. T., and T. S. Elias (eds.). 1977. *Extinction is Forever.* New York Botanical Garden, Bronx.

Prescott-Allen, R., and C. Prescott-Allen, 1982a. The case for in situ conservation of crop genetic resources. *Nature and Resources* 18:15-20.

———. 1982b. *Economic Contributions of Wild Plants and Animals to Developing Countries*. Report to the U.S. AID/MAB Program, Washington, D.C., February.

———. 1983. *Genes from the Wild: Using Wild Genetic Resources for Food and Raw Materials*. International Institute for Environment and Development, London.

Prest, J. 1981. *The Garden of Eden: The Botanic Garden and the Re-creation of Paradise*. Yale University Press, New Haven, Connecticut.

Purseglove, J. W. 1959. History and functions of botanic gardens with special reference to Singapore. *The Gardens' Bulletin* (Singapore) 17(2):125-154.

———. 1974. *Tropical Crops: Dicotyledons*. Wiley, New York.

———. 1975. *Tropical Crops: Monocotyledons*. Wiley, New York.

Rajapakse, H. 1984. The role of botanic gardens in Sri Lanka. Paper presented at the International Symposium on Botanic Gardens of the Tropics, Penang, Malaysia, 17-19 December.

Ramawas, S., and C. Nagai. 1984. Chromosome variability in progenies of selfed *Saccharum spontaneum*. *In*: D. J. Heinz and M. K. Carlson (eds.), *Annual Report 1983*. Hawaiian Sugar Planters' Association, Aiea, pp. 7-8.

Rastogi, K. B., and S. S. Saini. 1984. Inheritance of resistance to pea blight (*Ascochyta pinodella*) and induction of resistance in pea (*Pisum sativum* L.). *Euphytica* 33:9-11.

Raven, P. H. 1976. Ethics and attitudes. *In*: J. B. Simmons, R. I. Beyer, P. E. Brandham, G. L. Lucas, and V.T.H. Parry (eds.), *Conservation of Threatened Plants*. Plenum Press, New York, pp. 155-179.

———. 1981. Research in botanical gardens. *Bot. Jahrb. Syst.* 102(1-4):53-72.

Ribeyrolles, C. 1941. *Brasil Pitoresco*. Livraria Martins, São Paulo.

Rick, C. M. 1967. Exploiting species hybrids for vegetable improvement. *Proceedings of the XVIII International Horticultural Congress* 3:217-229.

———. 1973. Potential genetic resources in tomato species: clues from observations in native habitats. *In*: A. M. Srb (ed.), *Genes, Enzymes, and Populations*. Plenum Press, New York, pp. 255-269.

———. 1979. Potential improvement of tomatoes by controlled introgression of genes from wild species. *In*: A. C. Zeven and A. M. Van Harten (eds.), *Proceedings of the Conference: Broadening the Genetic Base of Crops, Wageningen, Netherlands, 3-7 July 1978*. PUDOC, Wageningen, pp. 167-173.

Rick, C. M., and P. G. Smith. 1953. Novel variations in tomato species hybrids. *American Naturalist* 87:359-373.

Rickett, H. W. 1956. The origin and growth of botanic gardens. *The Garden Journal of New York Botanical Garden* 6(5):133-135, 157-159.

Ridley, H. N. 1903. *Annual Report on the Botanic Gardens for the Year 1902*. Government Printing Office, Singapore.

———. 1907. *Annual Report on the Botanic Gardens Singapore and Penang for the Year 1906*. Kelly and Walsh, Ltd., Singapore.

———. 1910. The abolition of the botanic gardens of Penang. *Agricultural Bulletin of the Straits and Federated Malay States* 9(3):97-105.

———. 1911. *Annual Report on the Botanic Gardens, Singapore, for the Year 1910*. Government Printing Office, Singapore.

Riley, R., V. Chapman, and R. Johnson. 1968. Introduction of yellow rust resistance of *Aegilops comosa* into wheat by genetically homoeologous recombination. *Nature* 217:383-384.

Roberts, E. H. 1983. Seed preservation facilities and procedures for national centres of developing countries. *In: 1983 Rice Germplasm Conservation Workshop.* International Rice Research Institute/ International Board for Plant Genetic Resources, Los Baños, Philippines, pp. 71-76.

Roca, W. M., J. A. Rodriguez, G. Mafla, and J. Roa. 1984. *Procedures for Recovering Cassava Clones distributed in vitro.* Centro Internacional de Agricultura Tropical, Cali, Colombia.

Ross, H. 1979. Wild species and primitive cultivars as ancestors of potato varieties. *In*: A. C. Zeven and A. M. Van Harten (eds.), *Proceedings of the Conference: Broadening the Genetic Base of Crops, Wageningen, Netherlands, 3-7 July 1978.* PUDOC, Wageningen, pp. 237-245.

Ross, R. W., and P. R. Rowe. 1969. Utilizing the frost resistance of diploid Solanum species. *American Potato Journal* 46:5-13.

Rugendas, J. M. 1941. *Viagem Pitoresco através do Brasil.* Livraria Martins, São Paulo.

Rutger, J. N. 1983. Applications of induced and spontaneous mutation in rice breeding and genetics. *Advances in Agronomy* 36:383-413.

Ryerson, K. A. 1933. History and significance of the foreign plant introduction work of the United States Department of Agriculture. *Agricultural History* 7:110-128.

——. 1967. The history of plant exploration and introduction in the United States Department of Agriculture. *In: Proceedings of the International Symposium on Plant Introduction, Tegucigalpa, Honduras, November 30-December 2, 1966.* Escuela Agrícola Panamericana, Tegucigalpa, Honduras, pp. 1-19.

Saravia, L., and G. Lesino. 1983. Non-conventional energy seed storage facility for genetic conservation. International Board for Plant Genetic Resources, Rome, mimeo (ACPG:IBPGR/83/130).

Sastrapradja, D. S. 1982. Regional viewpoints on the theme of the keynote address: the Indonesian viewpoint. *In: The International Association of Botanic Gardens: Its Future Role.* Proceedings of the Ninth General Meeting and Conference of the International Association of Botanic Gardens, Canberra, 17-20 August, pp. 18-21.

Sastrapradja, D. S., and M. S. Prana. 1980. The viewpoint on conservation and the role of botanic gardens in conservation. *Buletin Kebun Raya* 4(6):175-182.

Sastrapradja, S., and T. A. Davis. 1983. The Bogor Botanic Garden—a plant paradise. *Hemisphere* 27(6):321-327.

Sastrapradja, S., and T. A. Davis. 1984. The Bogor Botanic Garden. *Garuda Magazine* 4(2):22-27.

Sastrapradja, S., D. S. Sastrapradja, and U. Soetisna, n.d. The botanic gardens of Indonesia: how the public make use of them. Manuscript.

Sauer, C. O. 1938. Theme of plant and animal destruction in economic history. *Journal of Farm Economics* 20:765-775.

———. 1969. *Agricultural Origins and Dispersals: The Domestication of Animals and Foodstuffs*. M.I.T. Press, Cambridge, Massachusetts.

Sauer, J. D. 1967. The grain amaranths and their relatives: a revised taxonomic and geographic survey. *Ann. Missouri Bot. Gard.* 54:103-137.

Schroder, J., and J. Schell. 1983. Applications of genetic engineering to plant and animal production. *In*: L. W. Shemilt (ed.), *Chemistry and World Food Supplies: The New Frontiers, Chemrawn II*. Pergamon Press, Oxford, pp. 563-568.

Schroeder, C. A. 1967. Avocado introduction in southern California. *In: Proceedings of the International Symposium on Plant Introduction, Tegucigalpa, Honduras, November 30-December 2, 1966*. Escuela Agrícola Panamericana, Tegucigalpa, Honduras, pp. 61-69.

Scowcroft, W. R. 1984. *Genetic Variability in Tissue Culture: Impact on Germplasm Conservation and Utilization*. International Board for Plant Genetic Resources, Rome.

Sharma, S. C. 1984. Botanic gardens of India, present status and future development. Paper presented at the International Symposium on Botanic Gardens of the Tropics, Penang, Malaysia, 17-19 December.

Shaw, T. 1976. Early crops in Africa. *In*: J. R. Harlan, J.M.J. De Wet, and A.B.L. Stemler (eds.), *Origins of African Plant Domestication*. Mouton, The Hague, pp. 107-153.

Shebeski, L. H. 1983. The application of wide crosses to plants. *In*: L. W. Shemilt (ed.), *Chemistry and World Food Supplies: The New Frontiers, Chemrawn II*. Pergamon Press, Oxford, pp. 569-576.

Sheng-ji, P. 1984. *Botanical Gardens in China*. University of Hawaii Press, Honolulu (Harold L. Lyon Arboretum Lecture Series 13).

Silva, A. R. 1976. Application of the genetic approach to wheat culture in Brazil. *In*: M. J. Wright (ed.), *Plant Adaptation to Mineral Stress in Problem Soils*. Cornell University Agricultural Experiment Station, Ithaca, New York, pp. 223-231.

Simmonds, N. W. 1979. *Principles of Crop Improvement*. Longman, London.

Singh, J. 1980. Co-ordinated maize improvement project—its organization and significant landmarks in recent years in the history of maize improvement. *In*: J. Singh (ed.), *Breeding, Production and Protection Methodologies of Maize in India*. Indian Agricultural Research Institute, New Delhi, pp. 14-32.

Smith, J.S.C. 1984. Genetic variability within U.S. hybrid maize: multivariate analysis of isozyme data. *Crop Science* 24:1041-1046.

Smith, N. 1983a. New genes from wild potatoes. *New Scientist* 98:558-565.

———. 1983b. Triticale: the birth of a new cereal. *New Scientist* 97:98-99.

———. 1984. Review of B. Weinstein, *The Amazon Rubber Boom, 1850-1920* (Stanford University Press, Stanford, California, 1983). *In: Hispanic American Historical Review* 64(3):589.

———. 1985. A plague on manioc. *Geographical Magazine* 57:539-540.

———. 1986. *Botanic Gardens and Germplasm Conservation*. University of Hawaii Press, Honolulu (Harold L. Lyon Arboretum Lecture Series).

Smith, R. D. 1984. The influence of collecting, harvesting and processing on the viability of seed. *In*: J. B. Dickie, S. Linington, and J. T. Williams (eds.), *Seed*

Management Techniques for Genebanks: A Report of a Workshop Held 6-9th July 1982 at the Royal Botanic Gardens, Kew, U.K. International Board for Plant Genetic Resources, Rome, pp. 42-82.

Smith, S. 1984. A plant breeders' perspective on genetic diversity: a reply to Pat Mooney's *The Law of the Seed. Diversity* 6:19-23.

Sola, F. de. 1967. Introduction. *In: Proceedings of the International Symposium on Plant Introduction, Tegucigalpa, Honduras, November 30-December 2, 1966.* Escuela Agrícola Panamericana, Tegucigalpa, Honduras, pp. v-vii.

Souza, P. F. 1945. The Brazilian forests. *In:* F. Verdoorn (ed.), *Plants and Plant Science in Latin America.* Chronica Botanica, Waltham, Massachusetts, pp. 111-119.

Sprague, E. W., and R. L. Paliwal. 1984. CIMMYT's maize improvement programme. *Outlook on Agriculture* 13:24-31.

Stalker, H. T. 1980. Utilization of wild species for crop improvement. *Advances in Agronomy* 33:111-147.

Steele, A. R. 1964. *Flowers for the King; The Expedition of Ruiz and Pavon and the Flora of Peru.* Duke Univeristy Press, Durham, North Carolina.

Sun, M. 1984a. The mystery of Florida's citrus canker. *Science* 226:322-323.

——. 1984b. Pests prevail despite pesticides. *Science* 226:1293.

——. 1985. Plants can be patented now. *Science* 230:363.

Suneson, C. A. 1960. Genetic diversity—a protection against plant diseases and insects. *Agronomy Journal* 52:319-321.

Swaminathan, M. S. 1982. Beyond IR36—rice research strategies for the 80s. Paper presented at the International Centers' Week, World Bank, Washington, D.C., 10 November.

——. 1983. *Genetic Conservation: Microbes to Man.* Presidential Address of the XV International Congress of Genetics, New Delhi, India, December 12-21.

——. 1984a. Rice. *Scientific American* 250:81-93.

——. 1984b. Agricultural production. *The Lancet* 8415:1329-1332.

Sylvain, P. G. 1955. Some observations on *Coffea arabica* L. in Ethiopia. *Turrialba* 5(1-2):37-53.

TAC. 1985. Report on the second external program and management review of the International Board for Plant Genetic Resources (IBPGR). Technical Advisory Committee, Consultative Group on International Agricultural Research, Food and Agricultural Organization of the United Nations, Rome.

Tachard, G. 1686. *Voyage de Siam des Pères Jésuites.* Arnould Seneuze and Daniel Horthemels, Paris.

Tay, C. S., Y. K. Hwang, and W. H. Kuo, 1984. *1984 Progress Report: Genetic Resources and Seed Unit.* Asian Vegetable Research and Development Center, Shanhua, Taiwan.

Terry, E. R., and S. K. Hahn. 1982. Increasing and stabilizing cassava and sweet-potato productivity by disease resistance and crop hygiene. *In: Root Crops in Eastern Africa: Proceedings of a Workshop Held in Kigali, Rwanda, 23-27 November 1980.* International Development Research Centre, Ottawa, pp. 47-52.

Theobald, W. L. 1982. A tropical garden for the nation. *International Plant Propagators' Society Combined Proceedings for 1982* 32:246-251.

Timothy, D. H. 1972. Plant germ plasm resources and utilization. *In*: M. T. Farvar and J. P. Milton (eds.), *The Careless Technology: Ecology and International Development*. The Natural History Press, New York.

Timothy, D. H., and M. M. Goodman. 1979. Germplasm preservation: the basis of future feast or famine; genetic resources of maize—an example. *In*: I. Rubenstein, R. L. Phillipps, C. E. Green, and B. G. Gengebach (eds.), *The Plant Seed: Development, Preservation, and Germination*. Academic Press, New York, pp. 171-200.

Tsitsin, N. V., and V. F. Lubimova, 1959. New species and forms of cereals derived from hybridization between wheat and couch grass. *American Naturalist* 93:181-191.

Tucker, W. 1984. Seeds of discord: the U.N. and scientists clash over control of plant genes. *Barron's* 64(30):26.

Turner, B. L. 1974. Prehistoric intensive agriculture in the Mayan lowlands. *Science* 185:118-124.

Ullstrup, A. J. 1972. The impacts of the southern corn leaf blight epidemics of 1970-1971. *Annual Review of Phytopathology* 10:37-50.

UNDP/IBPGR. 1984. *European Cooperative Programme for the Conservation and Exchange of Crop Genetic Resources, Forages Working Group: Report of a Working Group held at the Fodder Crops and Pasture Institute, Larissa, Greece, 7-9 February 1984*. United Nations Development Programme/International Board for Plant Genetic Resources, Rome.

USDA, 1971. *The National Program for Conservation of Crop Germ Plasm (A Progress Report on Federal/State Cooperation)*. U.S. Department of Agriculture, Washington, D.C.

Valls, J.F.M. 1985. Groundnut germplasm management in Brazil. *In: Proceedings of an International Workshop on Cytogenetics of Arachis, ICRISAT Center, Patancheru, India, 31 Oct.-2 Nov. 1983*. International Crops Research Institute for the Semi-Arid Tropics, Patancheru, India, pp. 43-45.

Van der Maesen, L.J.G. 1984. Seed storage, viability and rejuvenation. *In*: J. R. Witcombe and W. Erskine (eds.), *Genetic Resources and Their Exploitation*. Martinus Nijhoff/Dr. W. Junk, The Hague, pp. 13-22.

Van der Plank, J. E. 1963. *Plant Diseases: Epidemics and Control*. Academic Press, New York.

———. 1968. *Disease Resistance in Plants*. Academic Press, New York.

Van Gorkom, K. W. 1945. Chapters in the history of cinchona II: the introduction of cinchona into Java. *In*: P. Honig and F. Verdoorn (eds.), *Science and Scientists in the Netherlands Indies*. Board for the Netherlands Indies, Surinam and Curaçao, New York, pp. 182-187.

Van Leersum, P. 1945. Chapters in the history of cinchona, III: junghuhn and cinchona cultivation. *In*: P. Honig and V. Verdoorn (eds.), *Science and Scientists in the Netherlands Indies*. Board for the Netherlands Indies, Surinam and Curaçao, New York, pp. 190-193.

Vavilov, N. I. 1940. The new systematics of cultivated plants. *In*: J. Huxley (ed.), *The New Systematics*. Clarendon Press, Oxford, pp. 549-566.

———. 1949. *The Origin, Variation, Immunity and Breeding of Cultivated Plants*. Chronica Botanica Company, Waltham, Massachusetts.

———. 1957. *World Resources of Cereals, Leguminous Seed Crops and Flax, and Their Utilization in Plant Breeding*. Academy of Sciences of the U.S.S.R., Moscow.

Vergara, B. S. 1984. Adaptability and use of Indica varieties in high-latitude areas. Paper presented at the International Crop Science Symposium, October 17-20, Fukuoka, Japan.

Von Borstel, R. C., and K. Lesins. 1977. On germ plasm conservation with special reference to the genus *Medicago*. *In*: R. A. Muhammed, R. Aksel, and R. C. Von Borstel (eds.), *Genetic Diversity in Plants*. Plenum Press, New York, pp. 45-50.

Voon, P. K. 1976. *Western Rubber Planting Enterprise in Southeast Asia, 1876-1921*. Penerbit Universiti Malaya, Kuala Lumpur, Malaysia.

Walbot, V., and C. A. Cullis. 1985. Rapid genomic change in higher plants. *Ann. Rev. Plant Physiol.* 36:367-396.

Walsh, J. 1984. Seeds of dissension sprout at FAO: Third World nations vote change in system to conserve germ plasm over objections of industrial countries which fund the program. *Science* 223:147-148.

Watkins, J. V. 1976. Plant introduction gardens of the Caribbean. *In: Proceedings of the International Symposium on Plant Introduction, Tegucigalpa, Honduras, November 30-December 2, 1966*. Escuela Agrícola Panamericana, Tegucigalpa, Honduras, pp. 27-31.

Watson, I. A. 1970. The utilization of wild species in the breeding of cultivated crops resistant to plant pathogens. *In*: O. H. Frankel, E. Bennett, R. D. Brock, A. H. Bunting, J. R. Harlan, and E. Schreiner (eds.), *Genetic Resources in Plants—Their Exploration and Conservation*. F. A. Davis, Philadelphia, pp. 441-457.

———. 1979. The recognition and use in multilines of genes for specific resistance to rust. *Indian Journal of Genetics and Plant Breeding* 39:50-59.

Weinstein, B. 1983. *The Amazon Rubber Boom, 1850-1920*. Stanford University Press, Stanford, California.

Weissich, P. R. 1982. Text of illustrated paper describing Honolulu Botanic Garden's history, physical set up and goals. Paper presented at Longwood Gardens, Kennett Square, Pennsylvania, Spring.

Wetter, R. L. 1984. The application of plant tissue cultures to plant production: an overview. Report prepared for the International Development Research Centre, Ottawa.

Wilkes, H. G. 1972. Maize and its wild relatives. *Science* 177:1071-1077.

———. 1977a. The world's crop plant germplasm—an endangered resource. *Bulletin of the Atomic Scientists* 33(2):8-16.

———. 1977b. Hybridization of maize and teosinte in Mexico and Guatemala and the improvement of maize. *Economic Botany* 31:254-293.

——. 1983. Current status of crop plant germplasm. *CRC Critical Reviews in Plant Sciences* 1(2):133-181.

——. 1984. Germplasm conservation toward the year 2000: potential for new crops and enhancement of present crops. *In*: C. W. Yeatman, D. Kafton, and G. Wilkes (eds.), *Plant Genetic Resources: A Conservation Imperative*. American Association for the Advancement of Science, Washington, D.C., pp. 131-164.

——. 1985. Teosinte: the closest relative of maize. *Maydica* 30:209-223.

Williams, J. T. 1982. Genetic conservation of wild plants. *Nature and Resources* 18:14-15.

——. 1984a. A decade of crop genetic resource research. In: J.H.W. Holden and J. T. Williams (eds.), *Crop Genetic Resources: Conservation and Evaluation*. George Allen and Unwin, London, pp. 1-17.

——. 1984b. The international germplasm program of the International Board for Plant Genetic Resources. *In*: W. L. Brown, T. T. Chang, M. M. Goodman, and Q. Jones (eds.), *Conservation of Crop Germplasm—An International Perspective*. Crop Science Society of America, Madison, Wisconsin, pp. 21-25.

Withers, L. A. 1980. *Tissue Culture Storage for Genetic Conservation*. International Board for Plant Genetic Resources, Rome.

——. 1982. Storage of plant tissue cultures. *In*: L. A. Withers and J. T. Williams (eds.), *Crop Genetic Resources: The Conservation of Difficult Material*. IUBS, Paris, Ser. B42:49-82.

Witt, S. C. 1985. *Briefbook: Biotechnology and Genetic Diversity*. California Agricultural Lands Project, San Francisco.

Wolf, E. 1959. *Sons of the Shaking Earth*. University of Chicago Press, Chicago.

Wolf, E. C. 1985. Conserving biological diversity. *In*: L. R. Brown, E. C. Wolf, L. Starke, W. U. Chandler, C. Flavin, S. Postel, and C. Pollock, *State of the World: A Worldwatch Institute Report on Progress Toward a Sustainable Society*. W. W. Norton, New York, pp. 124-146.

Wolf, M. S., and J. A. Barrett. 1980. Can we lead the pathogen astray? *Plant Disease* 64:148-155.

Woolley, C. L. 1930. *The Sumerians*. Clarendon Press, Oxford.

Wycherley, P. R. 1959. The Singapore Botanic Gardens and rubber in Malaya. *The Gardens' Bulletin* (Singapore) 17(2):175-186.

Xuan, V., and N. V. Luat. 1983. Vietnam. *In: 1983 Rice Germplasm Conservation Workshop*. International Rice Research Institute/International Board for Plant Genetic Resources, Los Baños, Philippines, pp. 42-43.

Yngaard, F. 1983. A procedure for packing long-term storage seed. *Plant Genetic Resources Newsletter* (Food and Agriculture Organization/International Board for Plant Genetic Resources) 54:28-31.

Zohary, D. 1970. Centers of diversity and centers of origin. *In*: O. H. Frankel and E. Bennett (eds.), *Genetic Resources in Plants—Their Exploration and Conservation*. Blackwell Scientific Publications, Oxford, pp. 33-42.

INDEX

Italic page numbers are illustrations; n = note, t = table.

Macao, 48

Mahakam (rice), 153

Mahogany, 51

Maize Institute (Braga, Portugal), 119

Maize mosaic virus, 13

Maize, 5, 8, 10, 13, 37, 73, 84, 85, 87, 94, 99, 100, 102, 146, 157-69 passim; accessions, 119-20, 120t; hybrids, 14, 17, 23, 33, 39; mutants, 119; seed companies, 31; SR52, 35

Malagasy Republic, 184

Malaysia, 54, 80, 176, 183-84

Male sterility, genetic, 14, 103, 164

Maligaya Rice Research Institute and Training Center (Philippines), 180

Mango, 48, 51, 61

Manipulation of the crop environment, 21

Manure, 158

Maqqari, al-, 43

Marginal lands, 3, 122

Mariel boat lift, 15

Markham, Clements, 52

Martinique, 54

Marx, Karl, agricultural naïveté of, 37

Maturity, early, 103

Mauritius, 44, 57

Max Planck Institute (Federal Republic of Germany), 159

Mayans, 13

McClintock, Barbara, 100

Medicinal plants, 43, 192

Medina Azahara (Spain), Moorish garden at, 42

Medium-term storage of genetic material, 67, 76, 139

Melons, 7

Merion (bluegrass), 24

Mexico, 7, 10, 33, 54, 59, 85, 87, 119, 132, 145, 158, 188, 189

Meyer, Frank, 60, 64, 65, 66

Microbiological Research Centers, 193

Microorganisms, germplasm conservation of, 192

Microtechniques for germplasm manipulation, 165

Millet, 96; accessions, 122-23, 124t; finger, 159; kodo, 75; pearl, 37, 86, 122-23

Miramar 63 (wheat), 24

MIRCENS, 193

Missionaries, 64-65

Missouri Botanic Garden (St. Louis, U.S.A.), 55

Moghul gardens in India, 43

Monastic gardens, 42

Mongolia, 190

Monoclonal antibodies, 107

Monocots (monocotyledonous plants), 99, 99n, 108

Monocropping, 25

Monoculture, 9

Monogenic resistance, xiv, 23-24

Moorish gardens (medieval Spain), 42-43

Mosaic disease: cassava, see Cassava mosaic disease; in sugar cane, 165

Mothballing, 74

Multiline varieties, 24-26

Multinational corporations and the interests of small farmers, ix, 32-33

Multiplication: of germplasm, 73; of cell and tissue cultures, 101. See also Regeneration

Mutants, maize, 119

Mutation, induced, 102-103

National Institute of Agricultural Sciences (Tsukuba, Japan), 119

National Institute of Biology legume gene bank (Bogor, Indonesia), 82

National Seed Storage Laboratory (Fort Collins, U.S.A.), 68, 73, 84, 113, 119

Natural catastrophes, 11

Nature reserves, 95-96, 195-96

Navel orange, 51, 60

Nematodes, 88, 156, 158, 163, 167

Nepal, 146

Nepalese Agricultural Inputs Corporation, 31-32

Netherlands, 163. See also Holland

New Guinea, 59

New York Botanic Garden (U.S.A.), 55

New Zealand, 59

Nicaragua, 94

Nicholas V, Pope, 43

Nif genes, 100-101

Nigeria, 33, 54, 146

Nitrogen fixation, 100-101

Nordic gene bank (Sweden), 76, 83, 95, 191

North Africa, 123

LIBRARY OF CONGRESS CATALOGING-IN-PUBLICATION DATA

Gene banks and the world's food.
Bibliography: p.
Includes index.
1. Germplasm resources, Plant.   2. Gene banks, Plant.
I. Plucknett, Donald L., 1931-

SB123.3.G46   1987   631.5′2   86-42841
ISBN 0-691-08438-6